# THE KOALA
A NATURAL HISTORY

## Australian Natural History Series

Series Editor: *Professor Terry Dawson*

The function of this series of titles is to make accessible accurate scientific information, complemented by high quality illustrations, on a wide variety of Australian animals. Written and illustrated by trained researchers and scientists, they are intended for students and biologists at both secondary and tertiary levels and, in general, for readers with a serious interest in animals and the environment.

Interested authors should contact the New South Wales University Press or Professor Dawson, School of Zoology, University of New South Wales, PO Box 1 Kensington 2033.

Books planned in the series include the wombat, the lyrebird, the goanna, the crocodile, the possum, the kangaroo, the emu, and the cockatoo.

Books published include:
*The Platypus*, Tom Grant   Illustrated by Dominic Fanning
*Little Penguin: Fairy Penguins in Australia*, Colin Stahel & Rosemary Gales   Illustrated by Jane Burrell
*The Koala A Natural History*, Anthony Lee & Roger Martin
   Illustrated by Simon Ward

# THE KOALA
## A NATURAL HISTORY

Anthony Lee and Roger Martin
Illustrated by
Simon Ward and Anthony Lee

Published by
**NEW SOUTH WALES UNIVERSITY PRESS**
PO Box 1 Kensington NSW Australia 2033
Telephone (02) 697 3403

© Anthony Lee, Roger Martin 1988

This book is copyright. Apart from any fair dealing for the purpose of private study, research, criticism or review, as permitted under the Copyright Act, no part may be reproduced by any process without written permission from the publisher.

**National Library of Australia**
Cataloguing-in-Publication entry:

Lee, Anthony K. (Anthony Kingston), 1933–
The koala, a natural history.

Bibliography.
Includes index.
ISBN 0 86840 354 7.

1. Koalas. I. Martin, Roger William, 1946–
II. Ward, Simon, III. Title. (Series:
Australian natural history series).

599.2

Printed in Australia by Australian
Print Group, Maryborough, Victoria.

# CONTENTS

**PREFACE**   7

**CHAPTER ONE** THE KOALA   11

    1.1 What's in a Name?
    1.2 Distinguishing Characteristics
    1.3 History and Ancestry
    1.4 Clues to Ancestry
    1.5 Scientific Recognition

**CHAPTER TWO** DISTRIBUTION, HABITAT AND TREE PREFERENCES   23

    2.1 Shrinking Forests
    2.2 Koalas and Eucalypts
    2.3 Why is the Koala Fastidious?
    2.4 Other Aspects of Diet Preference

**CHAPTER THREE** INSIDE KOALAS   34

    3.1 Nourishment from Leaves
    3.2 Teeth and Mastication
    3.3 Digestion
    3.4 Age and Tooth Wear
    3.5 Energy and Nutrition
    3.6 Brain
    3.7 Reproductive System

**CHAPTER FOUR** REPRODUCTION AND LIFE HISTORY   51

    4.1 Breeding Season
    4.2 Oestrous Cycles, Gestation and Early Development
    4.3 Birth and Pouch Life
    4.4 Emergence
    4.5 Independence and First Reproduction
    4.6 Longevity and Mortality

## CHAPTER FIVE BEHAVIOUR   61

    5.1   Daily Cycle of Activity
    5.2   Individual Behaviour
    5.3   Social Behaviour
    5.4   Infant-Parent Behaviour
    5.5   Communication

## CHAPTER SIX PREHISTORY AND HISTORY   78

    6.1   Aborigines and the Koala
    6.2   Discovery and Exploitation by Europeans

## CHAPTER SEVEN CONSERVATION AND MANAGEMENT   85

    7.1   Disease
           *Chlamydia* and Chlamydiosis
    7.2   Alienation of Habitat
    7.3   Koalas and Urbanisation
    7.4   Management
    7.5   The Future

## BIBLIOGRAPHY   97

## INDEX   101

# PREFACE

In his preface to the *The Platypus*, Tom Grant tells us how he sought to buy a copy of Harry Burrell's book of the same name, published in 1927, and was astounded to learn from a bookseller that Burrell's book was still considered the most authoritative work on the platypus. This prompted Tom to write his account. Similar circumstances led to this book. Ambrose Pratt's *The Call of the Koala* was published in 1937, and although long out of print and laden with dubious opinion rather than fact, it remains the most substantial popular account of the biology of the koala. In a way, neglect of the koala is even more surprising than neglect of the platypus. No Australian animal evokes more sentiment, and none has aroused more public concern for its future, than the koala.

As with Tom Grant's account of the platypus, we have attempted to write a book which can be read by interested lay persons, students and professional biologists seeking to gain a general overview of the biology of the koala. This knowledge has increased substantially over the past ten years. We now have a clearer perception of the ancestry of today's koala, of its diet and digestion, of its reproduction and life history, of its brain structure and behaviour, and of its management and future. However, much of this information is locked away in articles which are not readily available to naturalists and has not previously been drawn together into a general synthesis. It is no wonder that the community retains many misconceptions about koalas. We still hear that the koala is threatened with extinction, is drugged as a consequence of feeding on the leaves of eucalypts, and can transmit a venereal disease to anyone foolish enough to handle them. It is hoped this book will destroy these misconceptions and arouse our curiosity in a fascinating animal about which we still have much to learn.

We owe much of the information we have presented here to Peter Mitchell and Kath Handasyde, who have allowed us to refer to their unpublished studies, and to Gordon Sanson, John Nelson and Mark Hindell. Their criticism has been invaluable. We also thank Bob Warneke and Phillip Reed for providing specific information; Pat Coates and Thea McConnell for their patience and assistance in preparing the manuscript; Bruce Fuhrer, Peter Fell, David Curl, Michael Coyne and Kath Handasyde for photographs and

## PREFACE

preparation of photographs from which some of the figures were drawn, and to Janet Doyle for her many suggestions on clarity and style of the manuscript. Our own research on koalas has been generously supported by the Ingram Trust, Phillip Island Shire Council, The World Wildlife Fund, the Fund for Animals, Conservation, Forests and Lands (Victoria), Australian National Parks Service and American Express.

A.L. and R.M.

A male koala sitting in a characteristic upright posture. The dark stain in the centre of the chest marks the position of the sternal gland.

# CHAPTER 1

## THE KOALA

### 1.1 WHAT'S IN A NAME?

In their history of the koala, Tom Iredale and Gilbert Whitley (1934) suggest that the common name 'Koala' was derived from an Aboriginal dialect of eastern New South Wales. Ronald Strahan (1978) lists cullewine, koolewong, colo, colah, koolah, kaola and koala as published dialectal variations of the name in that region, 'complicated by problems of transliteration and printers' errors' (p.3).

The early settlers referred to koalas as sloths, monkeys, bears, and even monkey bears, adopting the unfortunate practice of transposing the names of animals which were already familiar to Europeans to Australian lookalikes. The virtual absence of a tail, together with their stocky build and their relatively long legs, gives the koalas a bear-like appearance, and undoubtedly led to their being referred to as, 'koala bears', or, 'native bears', names which persist today. They are, of course, marsupials not bears.

### 1.2 DISTINGUISHING CHARACTERISTICS

Adult koalas weigh between 4 and 13.5 kilograms, and so are among the largest tree-dwelling (arboreal) mammals. Adult males are up to 50 per cent larger than adult females, and koalas in Victoria are larger than those in Queensland. In Victoria the average weight of females is 7.9 kilograms and males 11.8 kilograms, while in Queensland the corresponding averages are 5.1 and 6.5 kilograms. We have caught several males in Victoria weighing 13.5 kilograms and females weighing 11 kilograms.

The empathy the koala generates among humans appears to be due to its being one of the few mammals which has a face rather than a muzzle, a trait it shares with humans. The eyes of the koala are directed towards the front rather than to the sides, and the appearance of the face is futher enhanced by the relatively high forehead and the large rhinarium or nose pad, which extends backwards towards the eyes. This impact is made all the more appealing by the koalas' habit of sitting upright (Frontispiece), in contrast to the usual mammalian postures of standing on all fours, or lying on the side or with the legs tucked under the belly.

Another feature that lends to their unusual appearance is their eyes, which are small in comparison to the size of the head, and their slit pupils, which are vertical rather than horizontal as in other marsupials. The iris is a golden brown.

The fur of the koala in southern Australia is thick and woolly and is thicker and longer on the back than on the belly. Both the inside and outside of the ears are heavily furred. Koalas in northern Australia have a short coat and this gives them a naked appearance. The colour and pattern of the coat varies considerably between individuals and with age. The back is generally grey, sometimes interrupted with white patches on the rump, and is darker in southern than in northern populations. In the south the coat becomes very tawny with age. The tips of the ears, beneath the chin, the insides of the arms, and the chest are usually white, and the lower abdomen is usually chocolate brown, particularly in old females.

Climbing marsupials usually have a well-developed tail which can be used to grasp branches. The koala is an exception as it has only a rudimentary tail which is concealed in the fur and can only be seen by close inspection. When climbing, the koala relies upon its long and powerful limbs, relatively large hands and feet and long, sharp, recurved claws (Fig. 1.1). The palms of the very large hands have pads, the surfaces of which are granulated rather than ridged as in possums. This again is surprising as ridging is assumed to increase the ability of the pads to grip the limbs of the trees. The first two digits of the hand oppose the remaining three, giving the hand (termed a 'forcipate' hand), the appearance

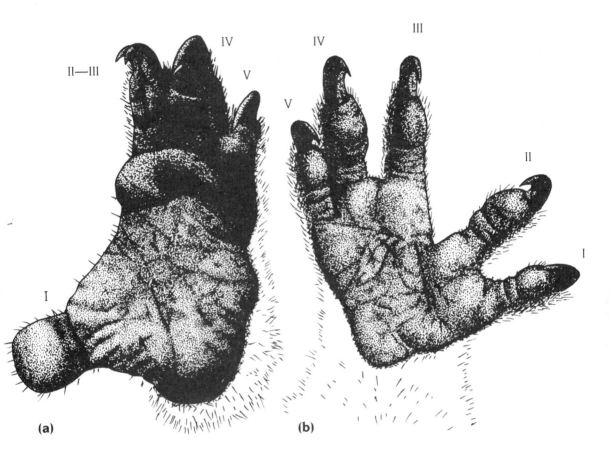

**Figure 1.1**
(**a**) A foot and (**b**) a hand of a koala. The first two digits of the hand oppose the remaining three in an arrangement referred to as a forcipate hand. All digits and all toes, except the first, are strongly clawed. The claws of the second and third toes form a comb. Unlike other arboreal marsupials, the pads on the hands and feet lack ridges.

of having three fingers and two thumbs, and a powerful grip. All digits possess claws. This arrangement of digits is also found in a number of the climbing possums. The structure of the foot, however, differs from that of other possums. It is short and broad and the big toe sticks out at right angles to the main line of the foot. This toe lacks a claw. The next two toes are long, are joined at the base, and lie close together to form a comb which they use to groom the fur. These and the remaining toes have prominent claws.

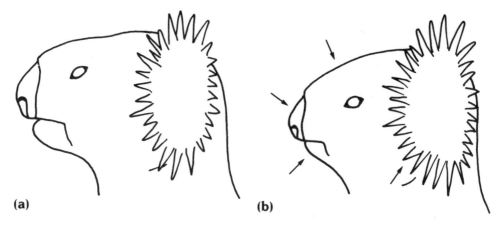

**Figure 1.2**
Profiles of the head of (**a**) a male and (**b**) a female koala, illustrating the distinctive 'roman nose' of the male and the straighter nose of the female.

Male koalas can be distinguished from females by the shape of the head (Fig.1.2). The head of adult males is relatively larger than that of females, and appears broader and squared-off in profile. Males also have a broad rather than pointed chin, and relatively small ears. Males possess a prominent gland in the centre of the chest, and the fur around this is usually orange-stained (Frontispiece). They also have a large pendulous scrotum which, as in other marsupials, is located in front of the penis. Females, of course, possess a pouch, but this is only conspicuous when occupied by a young. The pouch opens centrally and downwards when not occupied (Fig.1.3), and towards the back when there is a large young in the pouch. In this they differ from other arboreal marsupials which have an opening towards the anterior end of the pouch. The direction of the pouch opening in koalas seems an odd arrangement for an animal which spends most of its time sitting or climbing in an upright posture. Ronald Strahan tells the story of an old female koala in the Taronga Zoo which was unable to close the opening of the pouch and as a consequence lost young unless the pouch was closed with a clip!

## 1.3 HISTORY AND ANCESTRY

So what are the origins of this arboreal marsupial which in so many superficial features differs from other marsupials?
Sixty million years ago, the Australian, Antarctic and

South American continents were joined in a great southern land mass, Gondwanaland. It was a very different landscape from the Antarctica we know today; at least some parts were clothed in rainforest. Among the inhabitants of this land mass were the ancestors of the marsupials which are found in South America and Australia today, including the forebears of koalas. About 55 million years ago the Australian continent began to break away from Antarctica and move northwards. At this time there was no Antarctic ice sheet and the seas around Australia were warm. The winds that blew inland from these warm seas were laden with moisture and, as a result, heavy rainfall penetrated to the continent's heart. Rainforests prospered over much of the continent, but these contained a different mix of species from the tropical rainforests of northern Queensland today. Some of the trees in these ancient forests have relatives today in the highlands of New Guinea and others in the cool temperate rainforests of Tasmania. Present too were the ancestors of the modern eucalypts, acacias and casuarinas, trees which dominate most Australian forests today. Australian marsupials presumably diversified in these forests, but

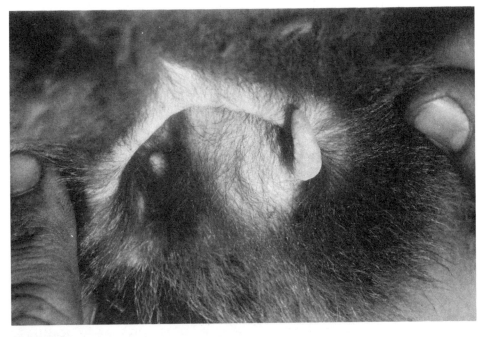

**Figure 1.3**
The backward opening pouch of a lactating koala. The elongated, suckled teat stretches beyond the entrance of the pouch.

unfortunately the fossil record from this time is so poor that it tells us nothing of this history.

Forty million years ago the Australian continent was still far to the south and about to lose its final connection with Antarctica. An Antarctic ice cap had appeared and the seas about Australia had cooled. The rainforests had by now taken on a distinctly cool temperate character and, although still widespread, had started to retreat in the face of declining rainfall. This trend towards aridity has continued to the present.

The oldest known fossil deposits in Australia which contain marsupials date from about 24 million years ago, but these fossils tell us little about the ancestry of koalas. For that we have to turn to deposits laid down about 15 million years ago when rainforests still persisted in central Australia. From these deposits in the now arid, almost treeless, north-east of South Australia, have come teeth, a skull and other fragments of bone of an animal that clearly bore affinities to the modern koala; but it is sufficiently distinct to be recognised as a distinct species, *Perikoala palankarinnica*. One thing that is very clear from these fossils is that the family to which fossil and modern koalas belong, the Phascolarctidae, was quite distinct from other families of marsupials by this time. The roots of the family are lost in that period of Australia's history which has yet to yield marsupial fossils.

The fossil record from 15 million years ago until the present reveals several different kinds of koalas. A single distinctive tooth, ascribed to a second species, *Litokoala kutjamarpensis*, has been obtained from another South Australian deposit estimated to date from about 10 million years ago. Younger still, at 4 to 4.5 million years, are six teeth from two sites in Queensland which have been assigned to two species, *Koobor notabilis* and *K. jimbarratti*. And finally a fragment of an upper jaw with a few teeth from limestone in south-eastern Queensland, about 50 000 years old, appears sufficiently similar to the modern koala to be included in the genus *Phascolarctos*, but as a distinct species, *P. stirtoni*.

Michael Archer (1978, 1984) has made an analysis of the relationships of these fossils based on the structure of the biting surfaces of the molar teeth (Fig. 1.4). Surprisingly he concluded that *L. kutjamarpensis* was closer to the ancestral stock than the older *P. palankarinnica*, and went on to predict that another species of *Litokoala*, will be found in fossil faunas older than 15 million years. If he is correct, then at least two lines of koalas occurred at this time. *Perikoala palankarinnica* was found to be closest to *Phascolarctos* and this stock was divergent from *Koobor*. So it seems that there may have been several different kinds of koalas living concurrently in Australia in the past. Unfortunately the analysis is based on few characters and little material, and so the conclusions are highly speculative. Perhaps the fabled Riversleigh deposits of north central Queensland will yield fossils which will allow Archer to test his hypothesis.

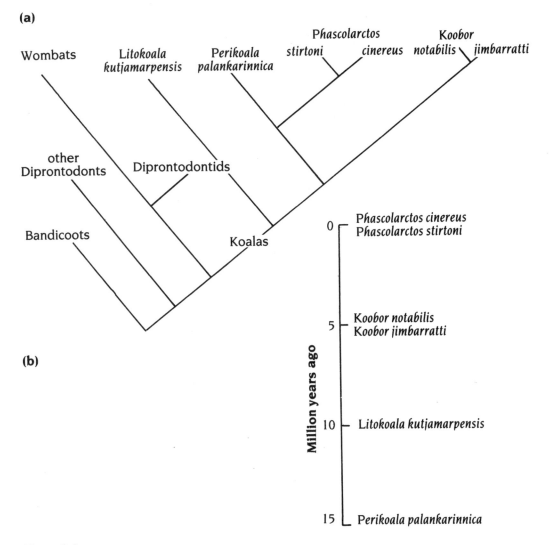

**Figure 1.4**
(a) The relationships of the koalas based upon an analysis of the structure of teeth (after Archer 1978). (b) The sequence of koalas in the fossil record. The analysis suggests that several types of koalas were contemporaneous in the past.

Since the fossil record provides few clues as to the origins of koalas and their relationships with other marsupials, biologists have had to turn to a comparison of the characteristics of living marsupials to provide these clues.

## 1.4 CLUES TO ANCESTRY

Although the koala has a number of characteristics which set it aside from other marsupials, there are nevertheless some traits which point to affinities with possums, kangaroos and wombats. One of these is the presence of three pairs of incisors in the upper jaw and a single pair of forward projecting (procumbent) incisors in the lower jaw. Another is the presence of comb (syndactyl) toes on the feet. The presence of both of these traits separates the diprotodont marsupials from other major groups of marsupials, the members of which lack either one or both traits. The order Diprotodonta incorporates possums, gliders, wallabies, kangaroos, wombats and koalas. This division of the diprotodonts from the rest of the marsupials seems to have occurred early in the history of marsupials.

The sorting out of the relationship of the koala to other diprotodonts has aroused some controversy. A small group of biologists has argued for a relationship with the ringtail possums on the grounds that they have similar molar teeth and similar hand structure to the koala. Others have come down in favour of an ancestry shared with the wombats, drawing attention to the many characteristics which are shared only by the koala and the wombat.

Oldfield Thomas (1888) first drew attention to the similarity of the crescent-shaped ridges which cap the biting surfaces of the molar teeth of koalas and ringtail possums, a condition referred to as selenodonty. This observation was taken further by Arthur Bensley (1903) who, on the basis of molar structure, divided the diprotodont marsupials into three groups. One of these groups contained the ringtail possum, the greater glider and the koala. Bensley's proposal to group the koala with the ringtail possum gained support from Frederic Wood-Jones (1924), who regarded their forcipate hands as a trait unique to this group of diprotodonts. Curiously their observations then escaped close scrutiny for about 40 years, even though alternative interpretations of the relationships of the koala had already been voiced.

The stimulus to re-examine the criteria for grouping the koala with ringtail possums came from studies of the structure of the sperm of marsupials, by Leon Hughes (1965), and the blood proteins (serology) of marsupials, by John Kirsch (1968). Hughes found that the sperm head of the koala was unusually hooked, and in shape and structure resembled the sperm heads of wombats rather than ringtail possums (Fig.1.5). Kirsch too found that the koala aligned with the wombat rather than the ringtail possum. This led Michael Archer (1976) to re-examine the structure of the molars of the koala and a variety of possums. He concluded that the selenodont condition in the koala and ringtail possums was similar to the ancestral condition, and was no basis for assuming a close and recent relationship between the koala and ringtail possum. Archer noted that the further back we go in the fossil record, the more common is the selenodont condition among diprotodonts. Retention of the selenodont condition among ringtail possums and koalas appear to be due to similarity in their diets rather than a recent common ancestry. This view is borne out by careful analysis of the molar structure, by Archer and by Ronald Strahan (1978), which shows that there are a number of significant structural differences between the grossly similar molar teeth.

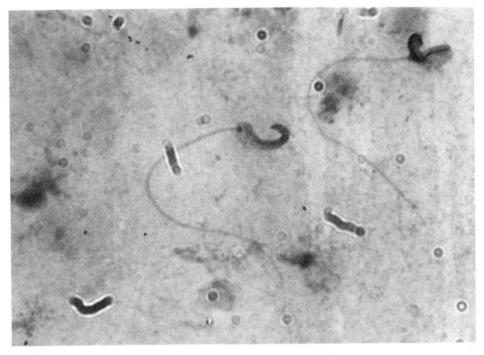

**Figure 1.5**
The strongly hooked head of a sperm of the koala, a character shared only with wombats.

The other characteristic which has been used to link the koala to the ringtail possum, the possession of a forcipate hand, has met a similar fate. Strahan noted that a forcipate hand was not confined to the ringtail possum, greater glider and the koala among arboreal marsupials, but also occurred in the cuscus. On other grounds the cuscus had been grouped with the brushtailed possum. Strahan also pointed out that there were subtle differences in the arrangement of bones that make up the hands of these animals. On this basis he suggested that forcipate hands had evolved on several separate occasions among diprotodont marsupials, and certainly independently in the koala and ringtail possum. So the hypothesis linking the koala to the ringtail possum has not withstood careful analysis.

It is interesting to note that the first description of the koala, published by E. Home in 1808, referred to the koala as another species of wombat, but gave no reason for this conclusion. The first formal recognition of the affinities between the koala and the wombat came from H. Winge (1893), who compared the structure of their skulls. Winge's conclusion was later supported by a detailed anatomical study of the entire animal by C.F. Sonntag (1921).

Among the external characters which are unique to the koala and wombat among the diprotodonts are a rudimentary tail (with a reduction in the number of vertebrae supporting the tail), granulated rather than ridged pads on the hands and feet, long rather than short syndactylous toes, a central-to-backward rather than forward-opening pouch, and one pair rather than two pairs of teats in the pouch. They also share a variety of internal characters which are not found in other diprotodonts, including cheek pouches, a 'gastric gland' in the stomach, similar gross structure of the liver and of the reproductive system, and a number of other characteristics of the skeleton and muscles.

The cheek pouches sit in the walls on either side of the mouth. Cheek pouches are common in rodents where they are often used to carry food from the site where it is gathered to caches near their nest. Koalas and wombats do not cache food and it is difficult to visualise the function of their cheek pouches, especially as they are small and only occasionally contain food. Perhaps they served some function in the common ancestor of the koala and wombat.

The so-called 'gastric gland' is a structure about 4 cm in diameter which sits in the wall of the stomach. It consists of about 25 crypts which open into the cavity of the stomach through a series of openings, reminiscent of the openings of a pepper pot. The gland probably secretes gastric juice, a mixture of digestive enzymes and acid. Wombats are the only other marsupials which possess a similar gland.

The derivation of the koala from a common ancestor with the wombat, especially a ground-dwelling (terrestrial) ancestor, provides a plausible explanation for the absence of a prominent tail. It is difficult to envisage circumstances favouring the loss of a tail had the koala been derived from an arboreal ancestor. Arboreal possums use their tail as another limb, often grasping branches with the tip while they manipulate food with both hands. Why would an arboreal animal give up this 'limb'? Likewise this ancestry would provide a plausible explanation for the absence of ridging on the pads of the hands and feet, and the direction of the pouch opening. The opening of the pouch in wombats, as in koalas, is displaced towards the back of the pouch as the young grows, and this may prevent the opening acting as a scoop when the distended pouch drags on the ground.

## 1.5 SCIENTIFIC RECOGNITION

Some recent classifications of the marsupials have recognised the affinities between the koala and the wombat by including them in a superfamily, the Vombatoidea. Within this superfamily the koala is placed in the family Phascolarctidae and the wombats, in the family Vombatoidea. Others avoid linking these families. This tells us that taxonomists view the gap between wombats and koalas as substantial.

The generic name of the only living species of koalas, *Phascolarctos*, is derived from the Greek *phaskolos* meaning pouch and *arktos* meaning bear. This name was proposed by the French naturalist H. de Blainville who described a specimen from New South Wales during a visit to London in 1814. The trivial name, *cinereus*, derived from the Latin *cinereus* meaning ash-coloured, we owe to a German biologist, H. Goldfuss, who used the name in a description published three years later. Other names and combinations of names have been proposed but these have all given way to *Phascolarctos cinereus*.

Early in this century, Oldfield Thomas (1923) established the subspecies *P. cinereus adjustus* for populations in Queensland, and Ellis Troughton (1935) then referred to Victorian populations as *P. cinereus victor*, leaving the populations in New South Wales as *P. cinereus cinereus*. These subspecies

# CHAPTER ONE

were distinguished on adult size, the shape of the muzzel, and pelage thickness and colour. However koalas, unlike the early federalists, do not cling to state boundaries, and it is now clear that these subspecies create artificial boundaries in a north-south trend or cline. As a consequence all populations are today referred to under the scientific name *Phascolarctos cinereus*.

# CHAPTER 2

# DISTRIBUTION, HABITAT AND TREE PREFERENCES

## 2.1 SHRINKING FORESTS

During the 15 million years that span the known fossil history of koalas, the face of Australia changed substantially. The forests that were once widespread in central Australia have retreated to the continent's margins, and in their place have developed communities of plants more tolerant of aridity. The composition of those forests has also changed. As far as the koala is concerned the most dramatic change may have occurred over the past 100 000 years. During this period the amount of charcoal in fossil sediments increased substantially, suggesting a marked increase in the frequency of fire. This was abetted during the last 50 000 years by the arrival of Aborigines who used fire to alter the vegetation to suit the game they hunted. Analyses of the kinds of pollen in the sediments that accumulated at the bottom of lakes during this period show that this change in the frequency of fire had a profound effect on the composition of plant communities, favouring species that were tolerant of fire. Among these were eucalypts and, as a result, we find today that eucalypts are the dominant trees in about 80 per cent of Australian forest communities. These changes in the vegetation probably contributed to the extinction of many browsing marsupials, so that today the widespread eucalypt forests support only four arboreal marsupials which feed primarily on foliage — the common brushtail possum, the common ringtail possum, the greater glider and the koala. All of these marsupials feed on the leaves of eucalypts and the koala and greater glider feed almost exclusively on eucalypt leaves.

These changes in the vegetation led to a substantial change in the distribution of koalas. We have already seen that phascolarctids once occurred in north-eastern South Australia where neither the koala nor forest occur today (Fig. 2.1). Clearly the geographic range of the family has contracted towards the edges of the continent as the climate became arid and the central Australian forests and woodlands gave way to desert shrublands. But this may not have been the sole cause of this contraction in range. Even as late as about 40 000 years ago, today's surviving species, *P. cinereus*, was found in the forested south-western corner of Western Australia. The cause of extinction of the koala in the south-west is unclear, but it may have resulted from hunting by Aborigines. By the time of the first European settlement the koala was confined to eastern Australia (Fig. 2.1). There is no evidence that the koala ever occurred in Tasmania.

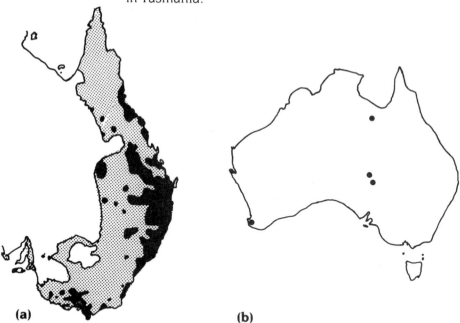

(a) (b)

**Figure 2.1**
(a) The probable distribution of koalas today (*shaded black*), and at the time of first settlement (*stippled*) (after *Strahan and Martin* 1982). (b) Sites from which fossil koalas have been taken which are beyond the boundaries of these distributions are indicated with a dot. Our knowledge of the current distribution of koalas will be greatly increased when the Australian National Parks and Wildlife Service/American Express survey is completed in 1988.

Further contraction of the range followed European settlement as forests and woodlands were cleared for agriculture, and as bushfires and hunting drastically reduced the numbers of koalas. Surveys in Queensland in 1967 and 1977 suggest that the range of the koala continues to shrink in that state. Today they are rarely seen in the forests of the south coast of New South Wales. Indigenous populations in the extreme south-east of South Australia are thought to have become extinct by the 1930s; however, they have been re-established in that state and over much of their former range in Victoria as a result of relocating koalas from vigorously expanding populations. Today they are found from southern Victoria in the south, to south-eastern South Australia in the west, and to the Atherton Tablelands in the north.

## 2.2 KOALAS AND EUCALYPTS

The association between koalas and eucalypts is not universal. There are substantial areas of forests dominated by eucalypts from which koalas are absent. They are most abundant in coastal woodlands, where they have been found at densities as high as ten animals per hectare, and in forests with open canopies. They are rare, or absent, in wet forests and in the south above 600 metres. Towards the inland margins of their range they are concentrated in trees fringing watercourses, and are thinly dispersed or absent from intervening woodlands.

Some of this variation in abundance appears to be due to climate, but a substantial proportion can be related to the types of eucalypts present. In southern Victoria and in South Australia, koalas are in greatest abundance in forests and woodlands in which manna gum (*Eucalyptus viminalis*) is the dominant tree species. Further north they favour communities in which forest red gum (*E. tereticornis*) or river red gum (*E. camaldulensis*) are the principal species. These are eucalypts which have a coastal or riverine distribution.

This preference of koalas for the foliage of certain eucalypt species deserves further comment as they have the reputation of being remarkably fastidious in diet. Most of the evidence supporting this conclusion has come from observations of koalas kept in captivity where they are provided with foliage. The most thorough early description of the koala's preferences for certain eucalypts was provided by David Fleay (1937) who, in the 1930s, observed the responses of koalas in the Melbourne Zoo to foliage from a variety of Victorian eucalypts. Fleay noted that rough barked coastal manna gum provided the most acceptable foliage, except for a six or seven week period in mid-winter. During this period the koalas spent most of the day and night huddled in the forks of trees, and showed extreme fastidiousness in their choice of food. At this time they still ate the foliage of long-leafed box (*E. goniocalyx*), and river red gum (*E. camaldulensis*) and forest red gum, but almost totally rejected manna gum. Fleay also noted that species such as narrow-leafed peppermint (*E. radiata*) and swamp gum (*E. ovata*) were eaten either sparingly or for short intervals during the year, and others such as sugar gum (*E. cladyocalyx*) were totally rejected.

Fleay's observations were prompted by concern that a high death rate among koalas in the Zoo was due to the provision of inappropriate foliage. This notion was championed by Ambrose Pratt (1937) who, in his popular account of the problem in *The Call of the Koala*, promoted the view that koalas are fastidious. However behaviour in captivity often gives a very false impression of behaviour under natural circumstances.

Two kinds of observations provide our knowledge of the natural diet of koalas. The first are records of chance observations of koalas feeding in the wild. These confirm that koalas show a preference for certain species in the genus *Eucalyptus*, but also show that the range of species eaten is much greater than the range they are usually offered in captivity. From this sort of information Bob Warneke assembled a list of 24 species of *Eucalyptus* in which koalas have been found in Victoria and identified nine as providing prime browse (Table 2.1). Koalas have also been observed feeding on species outside of the genus *Eucalyptus*, including golden wattle (*Acacia pycnantha*), hop wattle (*A. stricta*), cherry ballart (*Exocarpus cupressiformis*), pink box (*Tristania conferta*), swamp box (*T. suaveloens*), native kapok (*Bombax malabrica*), coast tea-tree (*Leptospermum laevigatum*), swamp paperbark (*Melaleuca ericifolia*), and the introduced apple (*Malus communis*) and Monterey pine (*Pinus radiata*). The discovery of their use of Monterey pine came as a surprise. Two students had been watching a group of koalas in an isolated grove of manna gum on French Island and noted that some animals disappeared from the eucalypts, sometimes for a week or more. These were ultimately discovered feeding in an adjacent stand of old pines. Koala faeces beneath the trees were packed with the remains of pine needles. We have subsequently found evidence of koalas feeding on Monterey pine on Phillip Island. Relating these observations to foresters seems to arouse their concern. Whether koalas can feed and survive indefinitely on these alternative food resources is not known. Our observations suggest that even where individuals persistently feed upon coastal tea-tree and swamp paperbark, eucalypt foliage (manna gum) consistently makes up the bulk of the diet.

The second kind of observation leads to a more systematic analyses of food preferences. This we owe to Michael Robbins and Eleanor Russell (1978), who found that the trees that koalas rest in during the day are those they either fed on the previous night or will feed upon the following night. Koalas usually feed in four to six bouts each 24 hours and tend to remain asleep in the same tree after they have fed. This is important, as it is easier to detect koalas during the day than to stumble around the forest searching for them at night. It allows us to obtain a substantial number of observations simply by walking through the forest, searching for koalas and identifying the trees they occupy. We have subsequently confirmed this relationship and found very few exceptions; koalas sometimes retreat to the dense foliage of acacias on hot days, but are rarely seen feeding on the foliage, and they are occasionally seen resting in dead trees.

Use of this relationship has provided a wealth of information on the types of trees used by koalas. In their study area just north of Sydney, Robbins and Russell found that the koalas preferred grey gum (*E. punctata*) and scribbly gum (*E. racemosa*) from the seven eucalypts available to them, and even showed a preference for individual trees within these species. On Phillip Island, in Western Port Bay, Victoria, Mark Hindell and colleagues (1985) found that some koalas showed a preference for manna gum, while others preferred swamp gum or Tasmanian blue gum (*E. globulus*). They also found that koalas show a preference for individual trees among these species. In a separate study which provides the most elaborate analysis to date, Mark Hindell (1984) found that the population in the Brisbane Ranges, west of Melbourne, showed an overall preference for manna gum, and a preference for manna gum in all seasons except summer, when both manna and swamp gum were equally preferred. Females in this population consistently preferred manna gum and swamp gum, but a group of males also showed a preference for red stringybark (*E. macrorhyncha*). Hindell suggested that these may be young males excluded from areas where manna and swamp gum were the predominant eucalypts. This Phillip Island population favoured large trees, and also demonstrated a preference for individual trees. Hindell went on to define the home ranges of 20 of the koalas by noting the particular trees they were using. Manna gum was the predominant species in the majority of these ranges and, not surprisingly, most of the animals with home ranges dominated by manna gum showed no preference for a particular species from those available within their home range. They used the trees of species in roughly the proportion that was available to them. However, eight of the koalas did show a preference and, in each case, the preferred species was uncommon within the animal's range. Two preferred manna gum, two swamp gum and two red stringybark.

It is clear from these studies that koalas show a preference for certain species of eucalypts in nature and that this preference largely determines where they live. It is also clear that this preference may change with season, although there is no evidence that this is accompanied by a shift in home range. In the Brisbane ranges, the preferred species, manna and swamp gum, were intermixed, and the koalas simply switched from one species to the other without altering their range. There is also evidence that individual koalas differ in the species they prefer, prefer some trees of a species over others, and prefer large rather than small trees.

These observations raise many intriguing questions. Why do koalas prefer the foliage of eucalypts, particular species of eucalypts and particular individuals of these species? Why do preferences change with season? Why do individuals differ in their preference? And at what stage of their lives are their preferences established? Most of these questions are at present unanswered, but we can eliminate some of the answers that have been proposed and point to some potentially profitable lines of investigation.

## 2.3 WHY IS THE KOALA FASTIDIOUS?

Although leaves are an abundant source of food, their nutritional quality is poor when compared to the other principal foods of animals. Leaves contain little protein and lots of difficult-to-digest fibre. They commonly contain toxic chemicals and substances which inhibit their digestion and are thought to serve as defences against attack by herbivores. Suggestions as to why koalas prefer foliage from certain species, and certain trees of those species, have proposed that koalas are either selecting foliage for its nutritional qualities or avoiding foliage containing defensive chemicals.

Ambrose Pratt was probably the first to offer an explanation for dietary choice. In the 1930s a high mortality rate among koalas in the Melbourne Zoo caused considerable concern. Pratt suggested that this mortality was due to the koalas eating foliage containing the highly toxic compound hydrocyanic acid (prussic acid). He attributed the rejection of foliage from manna gum by the koalas in mid-winter, and the simultaneous death of some koalas, to the presence of cyanogenic glycosides. These he suggested were converted to highly toxic hydrocyanic acid by an enzyme in the leaf tissue which is released when the leaves are masticated. Pratt based this conclusion on evidence of hydrocyanic acid in the foliage of manna gum collected near Sydney, and on reports of benzaldehyde, another compound derived from cyanogenic glycosides, in oils distilled from eucalypt foliage. However, subsequent tests of coastal manna gum which had been rejected by koalas failed to demonstrate the presence of cyanogenic glucosides. Recently, Eric Conn has completed a survey of manna gum from over much of its range and found cyanogenic glycosides in only one individual. Nevertheless the foliage of certain eucalypts (e.g. *E. cladyocalyx*) is cyanogenic, and this could be the basis of rejection of some eucalypts.

Eucalypt leaves are a rich source of oils (called essential oils), and the foliage of some species is commercially harvested and distilled for oils which have high bactericidal properties. In high concentrations, some of these oils may be toxic and Pratt suggested that foliage containing high concentrations of one of these, cineole, may have killed some koalas. He suggested that koalas selected foliage with a low oil content. Pratt went on to note that koalas in Queensland prefer different species to those in Victoria, and attributed this difference to selection for different essential oils, cineole in the north and phellandrene in the south. On flimsy evidence Pratt proposed that cineole, in less than lethal amounts, lowered heat production in an animal and therefore reduced the need to lose heat. This he saw as advantageous for animals in a hot climate. He also suggested, without evidence, that phellandrene raised the level of heat production, and therefore its consumption was advantageous in a cold climate. Pratt's hypotheses were all based upon early analyses of the essential oil content of eucalypt leaves by methods which are now known to be imprecise. Ian Southwell (1978) has subsequently analysed the essential oil content of leaf samples from 200 trees. Southwell chose trees which showed different degrees of browsing from localities scattered over much of the range of the koala. The only evidence in favour of Pratt's hypothesis was a higher cineole content in the foliage of heavily browsed trees from the north. However there were no obvious north-south trends in phellandrene concentration in heavily browsed trees, or in phellandrene or cineole concentrations when all trees were considered.

David Fleay also commented on the oil content of the foliage of trees favoured by Victorian koalas, and proposed that cineole was a vital constituent of preferred foliage. However this hypothesis was also based upon early analyses of oil content and, contrary to Fleay's observations, Southwell found that cineole was absent or in minute amounts in some heavily browsed foliage.

Koalas have considerable capacity to detoxify essential oils by conjugating them with glucuronic acid in the liver, forming compounds called glucuronides. These compounds are not toxic and are excreted in the urine or into the gut by way of the bile. Ian Eberhard and his colleagues (1976) found that on average, no more than 15 per cent of the oils in the leaves of grey gum eaten by koalas appeared in the faeces, and only trace amounts appeared in the urine. A portion of the ingested oils would have appeared as glucuronides, and the remainder, they suggested, may have been excreted from the lungs and skin. So it may be that these oils pose no particular problem. They also noted that the carbonyl content of foliage did not influence palatability, and was removed from the food in the gut and presumably detoxified. It is the carbonyl fraction which provides the germicidal properties of eucalyptus oil. This could pose a threat to the culture of bacteria in the caecum on which, as we will see, the koala may depend.

Recently attention has focused on some of the less volatile components of eucalypt leaves. For example, Mark Hindell identified a compound which he thought to be a wax or resin from the cuticle of the leaves. The presence of this wax appeared to be related to the rejection of foliage from certain species. However it was also clear that this was not the sole factor involved in choice as its concentration in the preferred species was similar to that in some species that were avoided.

The possibility that a volatile component of the leaves is involved in diet choice deserves further investigation. Koalas frequently sniff leaves at the tips of branches while feeding, and often reject one branch of foliage in favour of another. This suggests that choice may be made on the basis of some volatile component released by the leaves. Hydrocyanic acid, essential oils and leaf waxes and resins, all fall into this category.

Other attempts to identify a basis for preference have focused on the quality of eucalypt leaf as a source of nutrients and water. D.E. Ullrey and his colleagues (1981) have claimed that the foliage preferred by koalas in the San Diego Zoo had, among other characteristics, higher protein, phosphorus and potassium concentrations, and lower fibre, lignin, oil, calcium and iron concentrations, than rejected foliage, but they did not make comparisons between species. Mark Hindell analysed the protein, fibre, lignin and moisture content of leaves of different species of eucalypts in the Brisbane Ranges. Manna gum, the species preferred overall, had the lowest lignin content, and was among a group of species which had a high water content. However, there was no single characteristic among those he investigated which could account for the koala's choice of species.

So we are still far from understanding the basis of the koala's preference for certain eucalypts. Intuitively it seems that the basis of choice lies in the odour of leaves. However, identifying the volatile compound or mixture of compounds involved from the formidable range that is available will be a difficult task. As a first step it may be simplest to begin by eliminating nutritional factors as the basis of choice.

## 2.4 OTHER ASPECTS OF DIET PREFERENCE

Young of koalas have a long period of association with the mother once they leave the pouch. During this period they are weaned, and the leaves they first ingest are leaves they can reach while riding on the mother's back. Often they are seen feeding on the same bunch of leaves as their mother. If preference is established at this stage, then the young may acquire its preference from the mother's choice of trees.

Later events may change this preference. Female young tend to stay in the vicinity of their mother's range and so may be able to maintain their initial preferences. Males, however, disperse and may visit habitat where they may be forced to eat the foliage of other eucalypts. In maturity they move to areas occupied by females and so may be able to re-establish their early preferences.

Certain circumstances may lead to a change in the choice of food trees, and we have found no evidence that this affects their well-being. In recent years we have relocated several groups of koalas into different types of forest. In one

**Table 2.1.** *Eucalyptus* species commonly (\*\*) and occasionally (\*) used by koalas as food

| | | |
|---|---|---|
| E. amplifolia | cabbage gum | * |
| E. blakelyi | Blakely's red gum | * |
| E. botryoides | southern mahogany | * |
| E. camaldulensis | river red gum | ** |
| E. camphora | broad-leafed sallee | * |
| E. cambageana | Coowarra box | * |
| E. cinerea | Argyle apple | * |
| E. citriodora | lemon-scented gum | * |
| E. creba | narrow-leafed red ironbark | * |
| E. dalrympleana | mountain gum | * |
| E. drepanophylla | Queensland grey ironbark | * |
| E. globulus | Tasmanian blue gum | ** |
| E. goniocalyx | long-leafed box | ** |
| E. grandis | flooded gum | * |
| E. haemastoma | scribbly gum | * |
| E. largiflorens | black box | * |
| E. macrorhyncha | red stringybark | * |
| E. maculata | spotted gum | * |
| E. melliodora | yellow box | * |
| E. microcorys | tallowwood | * |
| E. nicholii | small-leafed peppermint | * |
| E. obliqua | messmate | * |
| E. ovata | swamp gum | ** |
| E. pilularis | blackbutt | * |
| E. populnea | poplar box | * |
| E. propinqua | small fruited grey gum | * |
| E. punctata | grey gum | ** |
| E. robusta | small mahogany | * |
| E. rubida | candle bark | * |
| E. tereticornis | forest red gum | ** |
| E. thozetiana | mountain yapungah | * |
| E. viminalis | manna gum | ** |

experiment we took animals from French Island, where they had the choice of manna and swamp gum, and released them into swamp gum in Lysterfield State Park, east of Melbourne. Here they had the choice of eight different species of *Eucalyptus*. All koalas immediately dispersed from the release site, and over the next two months, wandered to all corners of the Park. Some settled in natural forest and were found principally in mann gum and long-leafed box; others settled in a plantation of river red gum.

Recently we introduced nine koalas to an island not far from French Island. Manna gum was abundant at the site of release but once again the animals dispersed. Most settled in a part of the island where either swamp paperbark or coast tea-tree were the predominant tree species. Examination of faecal pellets showed that they were feeding on these species as well as manna gum. So it seems that, although individual koalas have a preference for the foliage of one or two species of *Eucalyptus*, and this may influence where they occur, it does not restrict the range of tree species upon which they may feed.

Knowledge of the range of species which can provide foliage suitable for maintaining koalas will be important in the future so that we can select habitat for koalas which have been removed from crowded populations. The knowledge that koalas will tolerate change in the species they feed upon, and will feed on species with which they are not familiar, greatly increases the range of suitable habitat. Knowing why they prefer certain species and reject others may prove invaluable in the development of an artificial diet, which could prove invaluable where koalas need to be kept in captivity.

# CHAPTER 3

## INSIDE KOALAS

### 3.1 NOURISHMENT FROM LEAVES

The principal food of koalas is leaves, although buds, fruit and even bark are occasionally eaten. Koalas strip young leaves from branches, apparently indiscriminately, but choose the mature leaves they eat. They grasp the branch, pull it close to the nose and sniff among the leaves before selecting one to eat.

As a source of nourishment, leaves can be viewed as consisting of two components: the contents of the cells and the walls surrounding those cells. These leaf components differ in the ease with which they can be digested and in their constituent nutrients. Cell contents are relatively easy to digest and contain most of the proteins and lipids in leaves and some of the carbohydrates. But to gain access to the cell contents, herbivores must first rupture the cell walls. Cell walls are difficult to digest and consist principally of complex carbohydrates such as cellulose.

Mammalian herbivores lack digestive enzymes capable of digesting cellulose, and rely upon bacteria to perform this task. Most of these herbivores have a large sac at some point along the digestive tract which houses a culture of bacteria. The herbivore provides the culture with masticated leaves and derives nutrients from the culture, either by directly absorbing the products of bacterial fermentation of cellulose, or by digesting bacteria which flow from the sac down the intestine.

Some mammals have the site of fermentation in front of

the stomach, such as the rumen of a cow. These mammals derive a substantial part of their nourishment from fermentation of cellulose. Other mammals have the fermentation chamber located beyond the stomach and small intestine, and derive most of their nourishment from the more easily digested cell contents.

It has been suggested that this arrangement enables the mammal to digest and absorb most of the cell contents before they reach the bacterial culture. It must be remembered that while the bacterial culture furnishes the mammal with food, it also requires food to maintain itself. This is lost to the mammal.

Because cellulose and related compounds in the cell wall are difficult and slow to digest, food containing a high proportion of these compounds loses its bulk slowly as it passes down the digestive tract. This means that the passage of food through the digestive tract of herbivores cannot be maintained at the high rate typical of the tracts of carnivorous mammals which are able to rapidly reduce the food mass through digestion. Cell walls may also contain lignin, and this compound is not only difficult, if not impossible, to digest, but also has the effect of inhibiting cellulose digestion. As a consequence of these constraints on the digestion of leaves, the rate at which herbivores derive nourishment is slow by comparison with other mammals.

There are a variety of ways herbivores can minimise the difficulties of obtaining nourishment from a diet of leaves. One of these is to select young leaves. The cell walls of young leaves are thin and contain little lignin, and so are easily ruptured and digested. Furthermore young leaves contain proportionally more cell contents than cell wall. By merely selecting young leaves koalas would increase the digestibility of their food.

However this may substantially increase the costs of foraging, as young leaves tend to be dispersed in trees at the tips of branches. Also, young leaves are not available in some seasons. The difficulties of foraging solely on young leaves may prevent koalas relying on this strategy.

Another way herbivores may increase the rate of digestion is by increasing the efficiency of mastication. Mastication should achieve two goals: rupturing of the cell walls to expose the cell contents, and fragmentation of the cell walls to expose them to bacterial attack and increase the rate at which food passes through the gut.

CHAPTER THREE

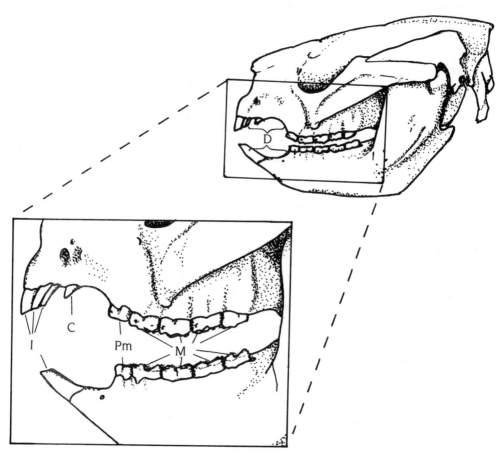

**Figure 3.1**
A lateral view of the skull of the koala showing the arrangement of teeth: incisors (**I**), canine (**C**), premolars (**Pm**) and molars (**M**). Leaves are fed through the diastema (**D**) between the incisors and premolars on one side of the mouth, across the premolars and molars on the other.

## 3.2 TEETH AND MASTICATION

The teeth of the koala, like the teeth of humans, are of four types: incisors, canines, premolars and molars. But they differ from human teeth in their shape, number and arrangement (Fig. 3.1). Of the three pairs of incisors in the upper jaw, the first pair are large and chisel-shaped, while in the remaining two pairs are small and peg-like, and sit hard against the first pair. The first pair of upper incisors oppose

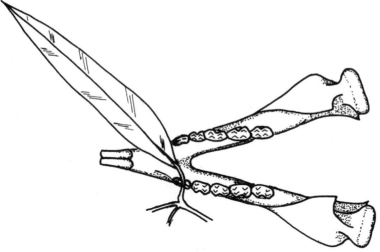

**Figure 3.2**
Position of a leaf on the lower jaw at the time it is removed from the branch. Closing the jaws forces the cutting blades of the premolars to shear past one another severing the leaf attachment.

the single pair of incisors in the lower jaw. These lower incisors are also large, and project forward and upwards in an arc. The opposing upper and lower incisors serve as a pair of pincers which are used together with one or other of the hands to grasp and manipulate stems as the koala feeds. They may also be used as a weapon when koalas fight one another.

A large gap or diastema follows the incisors and this is only interrupted by a pair of small, peg-like canines in the upper jaw. In each jaw the diastema ends at a pair of premolars. The blade-like premolars are each followed by four oblong molars. The premolars are used to strip leaves from branches. When suitable foliage is selected, the branch is steadied with a hand and a leaf is inserted obliquely into the mouth so that the leaf attachment enters the gap between the upper and lower premolars on one side (Figs. 3.2 and 3.3). The attachment is then severed by closing the jaws, forcing the cutting blades of the premolars to pass one another like the blades of shears. At the same time the koala may jerk its head upwards to pull the leaf free. Mastication of the leaf by the molars then begins.

The efficiency of the molars in mastication depends for a large part upon the structure of their biting surfaces. The biting surface of each molar is divided into quarters by two

# CHAPTER THREE

**Figure 3.3**
The position of a leaf between the lips of a koala at the time mastication is about to begin. The leaf is oriented by the lips and diastema on one side of the mouth so that it lies obliquely across the row of molars on the other side (*photograph by David Curl*).

valleys which run perpendicular to and intersect one another (Fig. 3.4). In each quarter there is a prominent crescent-shaped ridge, called a crista, which rises to a peak or cusp at roughly the midpoint of each ridge. During chewing, the lower jaw is moved from side to side as the mouth is opened and closed. The edges of the long curved cristae of the opposing upper and lower teeth mesh in a precisely maintained relationship, and are the only parts of the teeth that contact as the lower teeth are moved across the upper teeth. The leaf between the teeth is cut by the shearing action of the sharp leading edges of the cristae as they come into contact. The cristae of the upper and lower molars are curved in opposite directions. By virtue of this curvature, the cutting edges contact one another progressively at points, rather than along the entire crista at once. This results in all of the force of the bite being concentrated at the points of contact and enhances the efficiency of the cutting process.

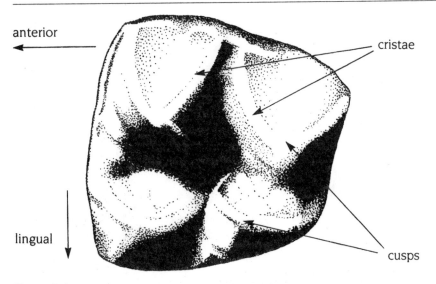

**Figure 3.4**
The biting or occlusal surface of the third upper left molar illustrating the crescent-shaped cristae rising to four prominent cusps.

When severed from the branch, the leaf lies diagonally across the mouth with the tip protruding through the diastema on one side, and the stem remnant lying across the row of molars on the other. The leaf is then drawn through the diastema by the lower molars on the opposite side as they drag across the upper molars in successive bites. At any one time only the rows of molars on one side of the mouth are used to masticate the leaf. The lips adjacent to the diastema on the opposite side act to guide the leaf across the molars.

The leaf moves one molar width with each chew, and between each chew it remains impaled on the cusps of the upper molars. The leaf is chewed four to nine times on one side of the jaw, and then swivelled with the tongue so that it protrudes through the opposite diastema. About of four to nine chews on one side of the mouth are always followed by four to nine chews on the other.

With narrow leaves, such as those of manna gum, the entire width of the leaf may be chewed with each bite. Usually the lateral and mid veins of the leaf remain intact so that the leaf retains its overall shape. Janet Lanyon and Gordon Sanson 1986), to whom we owe this description, likened the action of the molars to that of a double-edged pair of pinking scissors. The action of the molars on one side results in a series of serrated cuts which run obliquely across

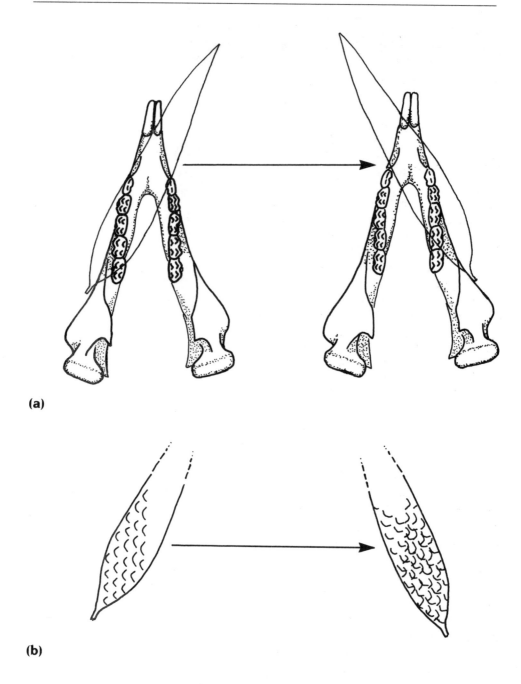

**Figure 3.5**
(a) The orientation of a leaf across the lower molars during the first (**left**) and second (**right**) series of bites. (b) The pattern of incisions across a leaf created by the cristae after each series of bites.

the leaf; the action of the molars on the other side superimposes a similar series of cuts perpendicular to the first, and in doing so cuts the leaf into minute particles (Fig. 3.5). As a consequence many of the cell walls are fragmented and the cell contents released. This meticulous preparation of the food in the mouth is the key to the koala's ability to extract nutrients from leaves.

## 3.3 DIGESTION

Earlier we mentioned the widely accepted view that there are two general ways mammalian herbivores tackle the problem of a diet rich in cellulose. Some herbivores have the culture of bacteria located before the stomach and are called 'foregut fermenters', and some have the culture located after the small intestine and are called 'hindgut fermenters'. The most notable feature of the gut of the koala is the enormous caecum, a spacious blind sac between 180 and 240 cm long, 10 to 15 cm in circumference, and with a capacity of 2 litres. The anatomist Collin MacKenzie (1918) judged this caecum to be proportionally the largest of any mammal, when differences in body size are taken into account. This caecum opens into the gut at the junction of the small intestine and the colon (Fig. 3.6). The presence of this large sac in an otherwise 'normal' gut suggests that the koala is a 'hindgut fermenter'.

We owe most of our knowledge of digestion in the koala to Steve Cork and his colleagues (1983), and their observations are surprising. Cork compared the composition of the ingested food (leaves of grey gum) with the composition of the faeces, to identify the components that are digested and absorbed by the animal. The comparison revealed that 69 per cent of the cell contents and only 25 per cent of the cell wall were digested. This evidence of reliance on cell contents is even more striking when it is realised that the leaves contain almost twice as much cell contents as cell wall. Cork went on to confirm the relative unimportance of cell wall as a source of nutrients for koalas by measuring the volatile fatty acid production in the caecum and in the large colon adjacent to the caecum (proximal colon). Volatile fatty acid is a product of bacterial fermentation of cellulose and is used by mammalian herbivores as a source of energy. Volatile fatty acid production in both the caecum and proximal colon of the koala was slow. The production of volatile fatty acid along the entire length of the gut only accounted for 9 per cent of the energy acquired by the animal, and only

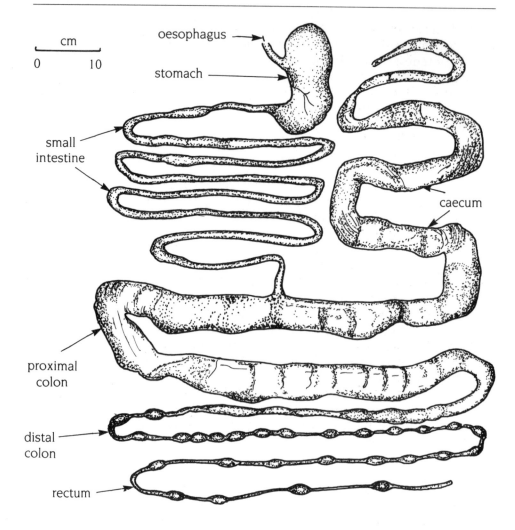

**Figure 3.6**
The gut of the koala arranged to illustrate the relationships and relative sizes of the component parts. The blind caecum, which opens into the proximal colon at its junction with the small intestine, is considered to be the largest in any animal when the size of the animal is taken into account.

two-thirds of this came from fermentation of cell wall. It seems clear from these observations that the unusually large caecum and proximal colon are serving a function additional to fermentation of cellulose. Unlike most other 'hindgut fermenters', the koala is clearly not relying on fermentation as a major source of nourishment. Nevertheless the modest supplement it gains from fermentation of cell wall may be important in its total energy budget.

A potential clue to the function of the caecum and proximal colon comes from study of the passage of materials down the digestive tract. Mastication and subsequent digestion of food in the stomach and small intestine produces digesta consisting of two components: a solution of exposed cell contents in which are suspended tiny particles of broken cell walls, and fragements of leaf material of different sizes. In the proximal colon the larger fragments are separated from the solution by a process which seems to involve settling of these fragments due to their weight. These are then compressed into faecal pellets in the distal colon and rectum, and then excreted. The solution and many of the small particles are retained in the caecum and proximal colon. Clearly the function of these organs is related to the retention of these components.

One suggestion is that the caecum and proximal colon are involved in the acquisition of nitrogen for production of proteins and nucleic acids. Leaves are generally a poor source of protein; eucalypt leaves contain only 4 to 15 per cent protein, suggesting that koalas may face a problem acquiring protein. Furthermore much of the protein ingested by the koala may be bound to tannins which are localised within the cell contents when the cells are intact, but mix with and bind to the protein when the cells are crushed during mastication. Tannin-binding inhibits the digestion of the protein in the stomach and small intestine. However bacteria in the caecum may be able to break this bond, releasing the protein for the animals' use.

Whatever the function of the caecum and colon, it is clear that separation and excretion of the large unproductive fragments in the digesta leads to a reduction of the volume of food in the gut. This will allow the koala to ingest additional food and so speed the rate at which it acquires nutrients.

## 3.4 AGE AND TOOTH WEAR

We have seen that the koala is largely dependent upon the cell contents as a source of nutrients and that the exposure and digestion of the cell contents requires efficient mastication of the food. However, thorough mastication of fibrous eucalypt leaves also causes abrasion to the biting surfaces of the teeth, and over time this alters their structure and their effectiveness in mastication.

For some time naturalists have been aware that the pattern of cutting edges on the surfaces of the molars changes with age and have used this as a means of determining the

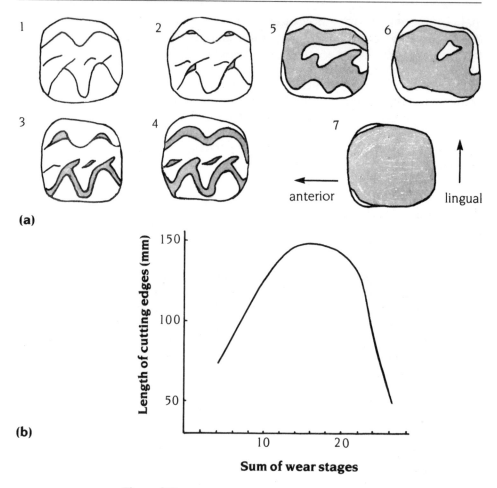

**Figure 3.7**
(a) Seven different wear stages of the occlusal surface of the first upper right molar of the koala. The solid lines represent the cutting edges and the stippled areas, exposed dentine. (b) The relationship between the combined lengths of the cutting edges of the four upper right molars and the sum of the wear stages (according to the above classification) of these teeth. As the teeth wear, the length of the cutting edges increases at first and then decreases.

age of animals in natural populations (Fig. 3.7). Janet Lanyon and Gordon Sanson (1986) have explored the consequences of these changes on mastication. Initially they measured the length of the cutting edges of the upper right molars at different stages of tooth wear and found that this increased at first and then decreased. By measuring the size of the fragments produced by animals with teeth at different stages of wear, they were able to show that the size of the fragments and the proportion of large fragments increased as the length of the cutting edges decreased. However, the size of fragments in the caecum did not change. They argued that animals with severely worn teeth, needing to maintain a supply of small fragments, do so by eating more. Old animals may thus find themselves in a 'Catch 22' situation: the need to eat more increases the rate of tooth wear, which increases the need to eat more. Ultimately they may not be able to meet their nutrient requirements.

## 3.5 ENERGY AND NUTRITION

Ken Nagy and Roger Martin (1985) watched a small group of koalas in late winter and observed that they each spent on average 14.5 hours of each day asleep, 4.8 hours resting but awake, 4.7 hours eating and 4 minutes travelling. Not a very energetic life! They then proceeded to measure the energy expenditure of the koalas while maintaining this lifestyle. The daily energy expenditure of these koalas was equivalent to 2090 kiljoules per animal, or about the energy content of 200 grams of breakfast cereal. They went on to estimate that this amount of energy could be obtained from 510 grams of eucalypt leaf and, as Cork has shown, this energy is mostly derived from the soluble carbohydrates and lipids of the cell contents. Nagy and Martin also estimated that the 510 grams of leaf would furnish twice the animal's daily nitrogen requirement, which suggests that nitrogen is unlikely to be a limiting resource for these animals, provided of course that tannins are not binding up the proteins in the leaves. Cork (1986) attributed the koala's success in satisfying its nitrogen requirements, while feeding on eucalypt leaves containing only 1.1 to 1.5 per cent nitrogen, to its ability to minimise nitrogen loss in the urine. Koalas appear to exceed other marsupials in this capacity. Surprisingly, Cork found that faecal nitrogen loss was relatively high. This he concluded was due, at least in part, to the sloughing of cells from the lining of the gut as a result of passage of their highly fibrous diet.

It is interesting to note at this point that Robert Degabriele and Terry Dawson (1979) found that the koala has a standard metabolic rate (the minimum amount of energy required to keep body systems functioning) which is only 70 per cent that predicted for a marsupial of its size. Furthermore, the standard metabolic rates of marsupials generally fall in the lower third of values for all mammals. It may be that the low energy and nitrogen requirements of koalas result from them rarely raising their metabolic rate below this low standard value.

Finally, Nagy and Martin also measured the water consumption of free-living koalas. Daily intake of the larger males at 475 millilitres per animal exceeded that of the smaller females at 358 millilitres per animal. In nature koalas rarely drink at bodies of free water, and in Nagy and Martin's study, the water was entirely derived from leaves and from water on the surface of leaves. Since males and females ate similar quantities of leaves, Nagy and Martin suggested that the males may have selected more succulent leaves or selected leaves wetted by rain.

Despite the monotony of their diet, we know of only one piece of evidence of koalas experiencing nutrient deficiency. In southern Victoria some populations of koalas live in areas where the leached sandy soils are deficient in copper. Koalas at these sites suffer an anaemia characteristic of animals feeding on a diet deficient in copper. Animals on French Island in Western Port Bay experience this anaemia. The only overt effect we have detected is their small body size as adults. Certainly it does not affect their fecundity. This is one of the most fecund populations we have discovered.

## 3.6 BRAIN

After watching the behaviour of koalas, it is not difficult to come away with the impression that they have a small brain. Much of their life is spent asleep, and when they are awake, their actions are usually made without haste. Indeed, measurements of the size of the brain show that it is only about 60 per cent of the weight and volume predicted for a marsupial of the koala's size. But koalas are not unique in possessing small brains. A number of folivorous marsupials, including some of the ringtail possums, the greater glider and the grey cuscus (*Phalanger orientalis*), are also small-brained.

What is exceptional about the koala is the degree to which the brain fills the brain case or endocranium. In most mammals the discrepancy between brain volume and the volume of the endocranium is less than 20 per cent, but John Haight and John Nelson (1987) have estimated that this discrepancy is a remarkable 61 per cent in the koala, due entirely to the koala's small brain! This does not mean that the koala's brain rattles loosely within the brain case.

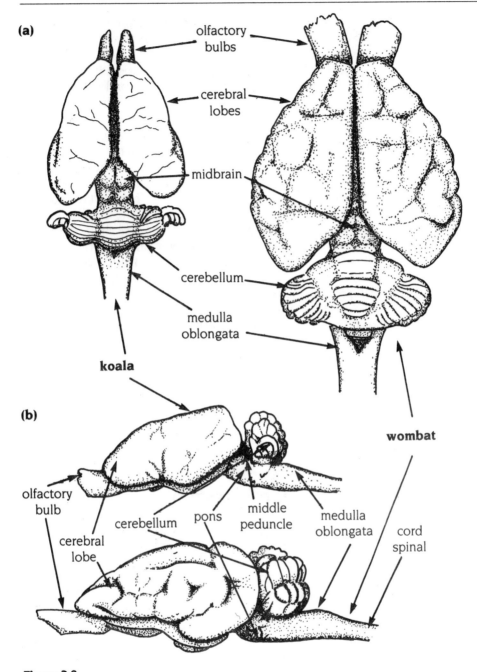

**Figure 3.8**
(**a**) Dorsal and (**b**) lateral views of the brains of a 9.7 kilogram koala and a 28 kilogram common wombat. In comparison to the wombat, the surfaces of the cerebral lobes of the koala lack folding, the cerebellum is small, and the spaces between the various lobes of the brain are large. Part, but not all, of the difference in the sizes of the brains can be attributed to their differences in body size.

Examination of the surface features of the cerebral lobes and cerebellum show that they closely fit the surface features of the brain case. Rather it is the spaces between the various lobes of the brain that are expanded. For example, there is a spacious gap behind the cerebral hemispheres through which the hind part of the forebrain (diencephalon) and the midbrain are clearly visible (Fig. 3.8).

Haight and Nelson's study of the anatomy of the brain of the koala reveals other peculiarities which tell something of the koala's behaviour. When compared with the brain of the common wombat (Fig. 3.8), the surface or cortex of the cerebral lobes lacks conspicious folding. This surface can be divided into areas which influence specific functions, and the associations between these areas and their functions are remarkably consistent between mammalian species. In the koala the areas associated with sight and hearing appear unexceptional. By contrast the area associated with skin sensation and with movement (parietofrontal cortex) is relatively small and thin, and has a cellular arrangement which suggests that it is poorly connected with other parts of the brain. This impression is reinforced by the small size of those regions of the brain (pons and middle peduncle) which house connections between the parietofrontal cortex and the cerebellum. The cerebellum, which coordinates movement, is also small. Together these observations suggest that the koala may be poorly endowed with motor behaviour and lack agility and fine control over its actions, a hypothesis we will examine later.

## 3.7 REPRODUCTIVE SYSTEM

The reproductive tract of the male has fairly typical anatomy for a marsupial, and in many respects this is also true of the female tract. One unusual feature in the female is the location of the ovaries. Usually the ovaries of mammals sit within the body cavity close to the openings to the fallopian tubes. Eggs shed by the ovaries must pass for a short distance through the body cavity before they are drawn into the fallopian tubes and begin their passage to the uterus. In the koala, each ovary is enclosed within a pouch called the ovarian bursa which is formed out of the mesenteries lining the body cavity and the mouth of a fallopian tube (Fig. 3.9). The only connection this pouch has to the body cavity is by way of a small pore in its wall. While this may seem an advantageous arrangement in that it ensures that the egg enters the oviduct, it may, as we will see, predispose koalas to

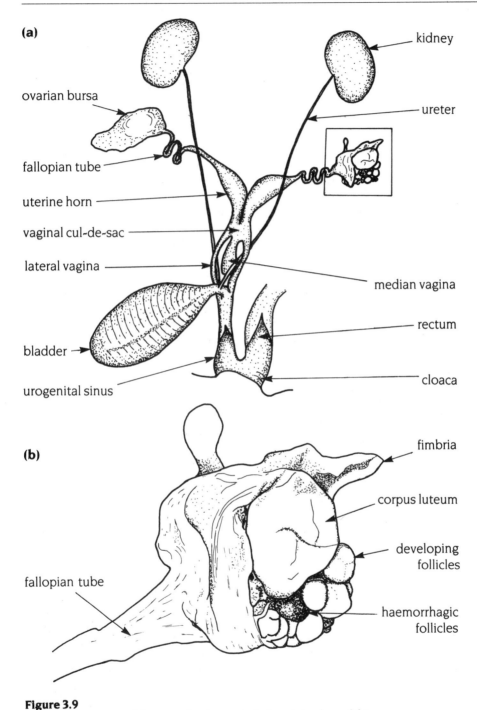

**Figure 3.9**
(**a**) The arrangement of the reproductive tract of a female koala, and (**b**) an enlargement of an ovarian bursa dissected to show the ovary within the folds of the bursae. The median vagina, joining the cul-de-sac to the urogenital sinus, may close after passage of a foetus in some individuals.

reproductive tract disease.

From each ovarian bursa a fallopian tube passes through four to six loops and ends in an arm or horn of the uterus. Each uterine horn begins with a thin-walled initial sector which has a glandular inner lining well supplied with blood vessels. It is in this sector of one of the two horns that the embryo will spend most of the period of gestation. The remainder of each uterine horn (cervix) has a thick muscular wall that acts as a valve, closing the uterus from the vaginal complex. The two uterine horns open independently into a thin-walled sac, the vaginal cul-de-sac, which may or may not be divided into two sides by a thin membrane. Two narrow lateral vaginae enter the outer walls of the cul-de-sac close to the cervix and terminate posteriorly in a relatively wide urogenital canal. In some animals a median vagina may join the cul-de-sac and the urogenital canal, but in others this is closed and may only open to allow the passage of the embryo before birth. Sperm probably reach the uterus by way of the lateral vaginae. This devious arrangement of the vaginae and of the uterus is a reminder that the forebears of marsupials, and indeed all mammals, had paired reproductive ducts. It is only in eutherians that we see a single vagina and in some eutherians, fusion of the uterine horns.

The bladder also opens into the urogenital canal which in turn joins the rectum at a common opening to the outside, the cloaca. This arrangement is typical of marsupials and the independence of the rectum and urogenital canal are only preserved by rings of muscle at their entrances. It is by way of the cloaca that the newborn koala departs for the pouch.

# CHAPTER 4

# REPRODUCTION AND LIFE HISTORY

## 4.1 BREEDING SEASON

Most marsupials breed seasonally and mate at a time which allows the young to emerge from the pouch when food is abundant and of the highest quality. The koala is no exception. Its young are weaned in late spring and early summer when eucalypts are flush with nutrient-rich new growth. A year elapses between the time a young is conceived and the time of weaning, and so koalas also mate in spring and summer.

Male koalas herald the onset of the mating season in September with an increase in bellowing. However the timing of breeding appears to be governed by the reproductive state of the females. Koalas are seasonally polyoestrous, that is, they have the potential for repeated oestrous cycles during the breeding season. The time of onset of oestrous cycles in spring depends upon whether or not the females are lactating. Kath Handasyde (1986) found that females which were not lactating when the breeding season began, first showed evidence of oestrous cycles in September. Some of these females mate and conceive in September, and give birth 35 days later in late October or November. However the majority of females are suckling a young in September, and Handasyde found that these females did not begin to show evidence of oestrous cycles until after this young was fully weaned. So although births may occur at any time between October and April, most births occur in December, January and February in southern Victoria. The distribution of births may be advanced by about a month in southern Queensland.

## 4.2 OESTROUS CYCLES, GESTATION AND EARLY DEVELOPMENT

Much of the early information on the reproductive biology of the koala was imprecise, and led to the suggestion that the koala was unusual among marsupials in having an oestrous cycle five to eight days shorter than the period of gestation. Kath Handasyde rejected this hypothesis when she found that the oestrous cycle had an average length of about 35 days, that is, roughly the same as the gestation length of 34 to 36 days.

It was also thought that females which lost a pouch young soon after birth, returned to oestrus and had the opportunity to conceive again in the same breeding season. If this were to occur, it would increase the probability of females breeding successfully each year. However Kath Handasyde found that in all of females that she examined which lost pouch young, a return to oestrous did not occur until the following season. As it is, few females lose their pouch young.

Finally, it had been suggested that koalas might exhibit embryonic diapause, a phenomenon found in certain kangaroos, wallabies and some small possums. In the red kangaroo (*Macropus rufus*) there is an oestrus shortly after a birth (postpartum oestrus). The fertilised egg which results from conception at this time, develops initially and then lies in a quiescent state (embryonic diapause) within the uterus until the preceding young is about to emerge from the pouch. The quiescent embryo then resumes development and is born shortly after the young leaves the pouch. Once again Kath Handasyde found no evidence of a postpartum oestrus or indeed oestrous cycles during lactation in the koala. She concluded that two phenomena inhibit the occurrence of oestrous cycles in koalas: decreasing day length in late summer and autumn which inhibits their occurrence in winter, and lactation.

We know only a little about the uterine life of the koala. Presumably, like other marsupials, gestation can be divided into two phases. In the first, the embryo develops into a hollow ball of cells, one cell thick, called a unilaminar blastocyst. This phase occupies at least two-thirds of gestation and during it, the embryo lies free within the uterus, enclosed by up to three membranes. The outermost of these membranes, the shell membrane, is the most persistent and completely separates the tissue of the embryo from the tissue lining the uterus. The second phase begins with the rupture of the shell membrane. During this phase the shape and organisation of the embryo changes dramatically, and a number of membranes develop out of the embryonic tissue. In most marsupials these membranes contact the uterine wall but do little else. In some, such as the koala, one membrane, the yolk sac, contacts and invades the tissue lining and uterus. The invasion is shallow and the area of the yolk sac involved is small and is not supplied with blood vessels. Two other membranes, the chorion and the allantois, which contribute to the placenta of eutherians and bandicoots, fuse and contact the uterine lining in the koala, but the chorion does not invade the lining and the allantois remains small. The role of these membranes as sites of exchange of materials between the embryo and mother has not been established in the koala.

## 4.3 BIRTH AND POUCH LIFE

Usually, koalas produce only a single young, but very occasionally twins are born. At birth the young (neonate) climbs from the cloaca to the pouch opening and then to

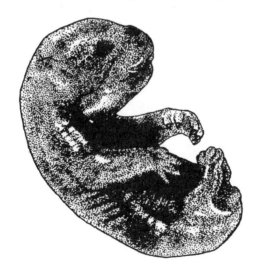

**Figure 4.1**
The pouch young of a koala at 1 week of age. The differences in development of the hands and feet which characterise the young at birth are still evident.

one of the two teats, securing the tip of the teat firmly between its lips. This young is about 19 millimetres in length and weighs approximately 0.5 grams or about one seventeen thousandth of the weight of the mother. As with other newly born marsupials, the young of the koala has well-developed forelimbs with digits equipped with minute claws, well-developed shoulder musculature and well-developed lips. By contrast, much of the rest of the body shows little resemblance to its adult appearance (Fig. 4.1). The hindlimbs, for example, are mere buds.

At seven weeks of age the young has a head length of about 26 millimetres. The head is large in proportion to the rest of the body, and the head and body are totally naked. The hindlimbs have developed and all toes except the first have claws. The nose pad, eyes, edges of the ears and the soles of the feet are darkly pigmented. The eyes are closed and bulbous and the ear canals still appear closed. The scrotum is visible in males. Pouch young essentially retain this appearance for the first 13 weeks of pouch life, and during this period remain firmly attached to the teat. By 13 weeks the young has attained a body weight of about 50 grams and a head length of 50 millimetres, but still bears little resemblance to the adult (Fig. 4.2).

## 4.4 EMERGENCE FROM POUCH

A transformation in appearance of the young takes place in the following weeks. Fine dark brown hair appears on top of the head and over the shoulders and back, and at about 22 weeks of age the eyes open. About this time, the young begins to poke its head out of the pouch for the first time.

This also marks a change in feeding behaviour. Over the next six to eight weeks the mother produces two kinds of excreta from the anus. Besides the typical firm faecal pellets, she produces a material call 'pap', which is soft and fluid. Pap is assumed to be derived from the contents of the caecum, and is eaten by the young. Pap is thought to prime the caecum of the young with bacteria which are essential for its function, but it may also provide the young with a rich source of bacterial protein. Feeding on pap marks the onset of a period of accelerated growth (Fig. 4.3).

At first, the young feeds on pap with only the head and shoulders protruding from the pouch entrance, but later it feeds clinging to the fur of the belly of the mother. There is some difference of opinion as to how young encourage the production of pap. Keith Minchin (1937), who was one of the

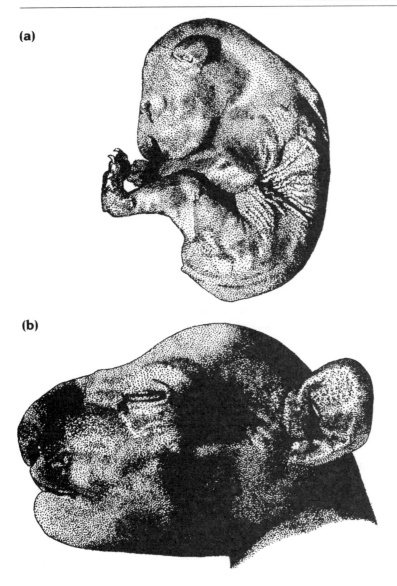

**Figure 4.2**
(**a**) The pouch young of a koala at 13 weeks of age. (**b**) A close view of the head. The young still bears little resemblance to the adult.

first to describe this phenomenon, suggested that the young pushed its head into the anus of the mother and used its forepaws to stretch the anus. However Malcolm Smith (1979) found that the young simply sweeps its mouth back and forth across the anus and occasionally waves its forelimbs, but never pushes against the belly of the mother.

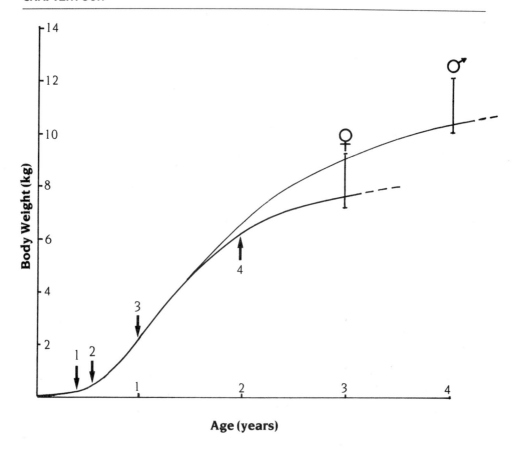

**Figure 4.3**
Growth in body weight of male and female koalas on French Island, Victoria. Numbers indicate when the head first appears out of the pouch (**1**), permanent vacation of the pouch (**2**), independence of the mother (**3**), earliest reproduction in females (**4**). Vertical bars depict the range of body weights. Males continue to grow until at least 4 years of age.

Minchin observed young feeding on pap for about an hour every second or third day, but it may occur more frequently as he did not watch his animals at night.

By 24 weeks of age the cub is fully furred and the first teeth erupt. These are the lower incisors and their eruption is soon followed by the appearance of three pairs of upper incisors. Initially all upper incisors contact the lower incisors and together they form a biting surface which enables the cub to chew material that is harder than pap. Curiously Malcolm Smith did not observe the cubs nibbling at leaves until they were at least 31 weeks of age, but we suspect that this may begin earlier. Later the second and third pairs of upper incisors lose contact with the lower incisors, but they remain, although apparently functionless, throughout adult life.

A period of about four to five weeks elapses between the time the head first protrudes from the pouch and the time the cub first emerges from the pouch. On emergence the cub weighs about 250 grams.

At 30 weeks the cub weighs about 500 grams and has a head length of 70 millimetres. It now spends most of the time out of the pouch clinging to the mother's belly. The fur has now changed from chocolate brown to the tawny grey typical of adults, and the first pair of molars has begun to erupt.

Some six weeks later the cub weighs 1 kilogram and no longer enters the pouch. It spends much of the time sitting on the mother's back, but returns to the mother's belly in cold, wet weather and to sleep. The second pair of molars has now erupted and the cub is becoming increasingly dependent upon eucalypt leaves for food. Nevertheless it still occasionally suckles. By now the suckled teat has elongated to the point where it may protrude from the opening of the pouch.

The cub encounters leaves while riding on the mother's back, and begins to bite at the edges of leaves without bothering to steady the leaves with the hands. Almost all leaves are smelt before being eaten and usually only the tips are ingested. Over the ensuing months this behaviour changes. The proportion of leaves they sniff declines, they begin to steady leaves with the forepaws before biting, and bite off whole leaves rather than parts of leaves.

Malcolm Smith did not observe cubs moving from contact with the mother until 37 weeks of age, and then the excursions were brief and quickly terminated if the mother moved. Often these cubs seemed to vacillate between moving further away and returning to the mother. Most early excursions involved walking along a branch, climbing and sniffing. Even cubs as old as 44 weeks ventured less than a metre from their mother, but by 48 weeks of age they were clearly more adventurous and no longer squeaked when the mother was removed. At this age mother and cub are often seen sleeping back to back.

Some of the behaviour of cubs during these early excursions appears to be rudimentary play. John Byers recently described to the authors behaviour in which cubs made repeated excursions from the mother and systematically attempted to reach across gaps between branches, increasing the gap with each attempt. John did not observe any play which appeared to be directed towards development of muscles or the capacities of the heart and lungs, and speculated that play in koalas (and other marsupials) may be principally directed towards development of the brain. If this is so, it is not surprising that koalas, with relatively small brains play little, whereas wombats, with relatively large brains, play frequently.

The cub remains with the mother until about 12 months of age when it weighs a little over 2 kilograms and has acquired the third pair of molars. If another young is born about this time, the bond between the yearling and mother abruptly breaks down. Mothers with newborn young strongly resist handling, particularly attempts to inspect the pouch. Yearlings attempting to suckle are treated to a cuff for their trouble, and although frequently found in the same tree as the mother, they are no longer tolerated upon her back. These yearlings are sometimes found with a surrogate mother, an adult female without a back young, or with a male courting the mother.

## 4.5 INDEPENDENCE AND FIRST REPRODUCTION

Young females stay in the vicinity of the mother for another year and some may settle in a home range nearby. These females may be mated by their father in subsequent breeding seasons. Young males usually disperse from their mother's range at about two years of age and may roam for the next two to three years before settling.

Females occasionally have their first young when about 18 months of age, when they approach adult size (Fig.4.8), but this young rarely survives pouch life. Most females breed first towards the end of their second year and may produce one young each year up until 10 to 15 years of age. Contrary to popular opinion we have found no evidence that females breed every second year.

Males are capable of reproducing at two years of age at the time of their dispersal from their mother's range, but most are prevented from gaining access to females by older and larger males. A two-year-old male weighing between 5.5 and 7 kilograms is simply unable to hold sway over an older male at 11.5 to 13.5 kilograms. Evidence that they are occasionally successful comes from a two-year-old male in the Brisbane Ranges population which showed evidence of a chlamydial infection. The organism responsible for this infection is transmitted venereally!

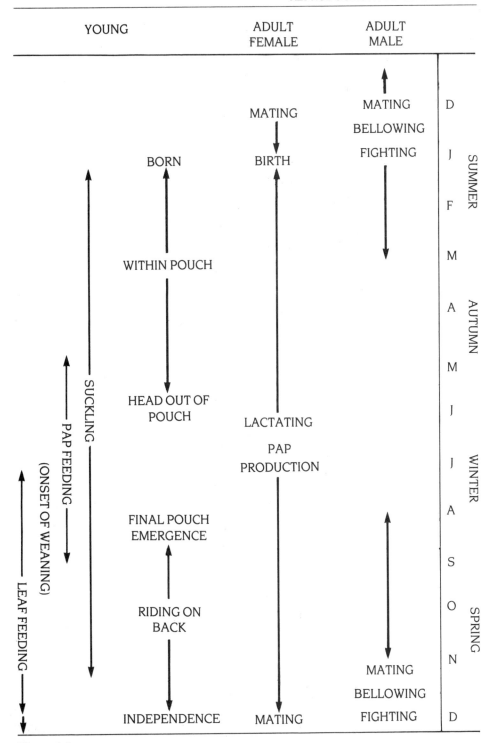

**Figure 4.4**
A year in the life of a koala.

## 4.6 LONGEVITY AND MORTALITY

There are a few records of longevity in nature. A female tagged on French Island was still breeding at ten years of age and a male at Walkerville, Victoria, was estimated to have died when 16 years old. A female in San Diego Zoo lived to 18 years, but it is probable that natural longevity is shorter than this, particularly for males.

In the past, the major causes of mortality appear to have been predation by Aborigines and dingoes (Chapter 6). Aborigines delighted in their flesh and, as we will see, appear to have had a profound influence on their numbers. There are also descriptions of dingoes pouncing upon koalas as they moved on the ground between trees. The only other known natural predator is the powerful owl (*Ninox strenua*), which takes young weighing less than 1 kilogram.

Bush fires and droughts may have also taken their toll. Koalas have no means of escaping fires which sweep through the crowns of eucalypts, and the few that survive these fires have little hope of avoiding starvation before the trees produce epicormic growth. In the early 1980s, a severe drought in central Queensland caused browning and loss of leaves from eucalypts, and resulted in substantial mortality among koalas.

Today, longevity in populations which escape the effects of fire and drought is probably determined by the incidence of disease or the rate of wear of their teeth. We will consider the impact of disease in Chapter 7. Those animals which escape debilitating disease may ultimately succumb from an inability to masticate sufficient food to meet their nutritional needs (Chapter 3). In winter, we occasionally find animals with severely worn teeth sitting at the base of a tree seemingly too weak to climb. They rarely, if ever, survive such an episode.

# CHAPTER 5

# BEHAVIOUR

## 5.1 DAILY CYCLE OF ACTIVITY

Watching koalas during the day gives the impression that they spend most of their lives resting or asleep, stirring occasionally to feed on eucalypt leaves. The reality of this impression is borne out by long-term observations of their behaviour which have consistently shown that about 19 hours of each day is spent resting or asleep, one to three hours is spent feeding, and the remaining small fraction is divided between moving between branches or trees, grooming and social behaviour (Fig. 5.1).

Feeding tends to be confined to four to six bouts of from 20 minutes to 2 hours' duration. These bouts can occur at any time of the day or night, but there is a tendency for koalas to feed in the hours immediately preceding and following dusk. For example, Michael Robbins and Eleanor Russell (1978) found that 66 per cent of their observations of feeding occurred during the night, and that observations of feeding were most common between 1600 and 2000 hours and least frequent between 1100 and 1400 hours. Lynda Sharpe (1980) identified three times when feeding was most frequent: one 2.5 hours before sunset, one 2.5 hours after sunset and a third 3 hours before dawn. She observed no feeding in the middle of the night.

Activities such as grooming, social behaviour and moving between trees, which occupy only a fraction of the daily cycle, also occur both day and night. But whereas social

# CHAPTER FIVE

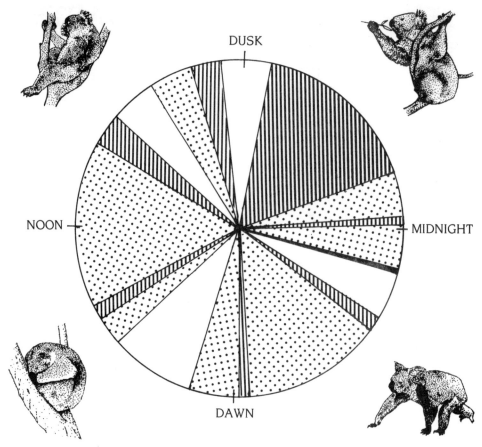

**Figure 5.1**
The daily cycle of activity of a koala. Vertical hatched areas signify periods of feeding; stippled areas, periods of sleeping; unshaded areas, periods of resting; and areas shaded black periods, moving between trees (*data from Nagy and Martin 1985*).

behaviour and moving between trees occur most frequently at night, grooming, according to Sharpe, is largely a daytime activity.

In the most comprehensive of all of these studies, Peter Mitchell (1988) found that 77 per cent of observations of feeding, 74 per cent of observations of moving and 73 per cent of social interactions occurred during the night, while 55 per cent of observations of resting and sleeping occurred during the day. Clearly, the daily cycle of activity of the koala is similar to that of some of the other large herbivorous marsupials, such as the large kangaroos, which although primarily nocturnal, exhibit some activity during the day.

## 5.2 INDIVIDUAL BEHAVIOUR

Koalas usually rest and sleep in the forks of trees, but are occasionally found stretched along a branch. In this respect they differ from most of the possums which either sleep in tree hollows or in nests of sticks and leaves. When sleeping in forks, koalas invariably face the trunk with their rump pressed against the limb. The precise sleeping posture is influenced by the weather. On hot days the limbs are extended and often lie free on either side of the trunk. The animal may recline along the limb and the head is held free of the chest, exposing the belly. On cold, wet and windy days they sit with their backs to the wind, with their arms folded against the chest and legs drawn against the belly. The shoulders are bent forward, the head is held against the chest and the ears are folded forward against the head.

These changes in posture probably have an important role in the regulation of body temperature. The compact posture of animals on cold days should serve to reduce heat loss by minimising the body surface exposed to the environment. The extended posture on hot days should favour heat loss. It is interesting to note here that Robert Degabriele and Terry Dawson (1979) found that the insulation provided by the back fur of koalas was superior to any yet measured for a marsupial, and was not greatly disturbed by wind. The belly fur, on the other hand, had only half the density of hairs which characterised the back fur and provided correspondingly less insulation. The paler belly fur also showed a greater capacity to reflect heat. Thus by exposing the belly on hot days, koalas may facilitate heat loss, and by shielding the belly from the weather on cold days, koalas may reduce heat loss. Koalas seem able to tolerate short periods of cold weather, and spend less time active and feeding under these conditions. However their absence from the colder mountain forests suggests a low tolerance of protracted cold weather. Surprisingly, they are less tolerant of hot weather. On hot days they seek shade on the ground, in trees with dense foliage or in mistletoe.

Koalas are primarily arboreal and only come to the ground to move between trees or seek shade. On the ground they usually walk slowly by moving diagonally opposite limbs in unison. The weight of the body is borne by the extended digits and the palm adjacent to the digits, and by the entire sole and the extended toes (Fig.5.2). Koalas are also capable of running by extending the forelimbs and then the hindlegs in a bounding gait, and may even gain sufficient momentum to lift all feet off the ground.

# CHAPTER FIVE

**Figure 5.2**
A koala walking on the ground. The weight of the body is borne on the extended digits and adjacent palm, and on the entire sole and extended toes (*photograph by David Curl*).

for the next spring. The next sequence is then repeated. Vital to their climbing ability are the long, recurved claws which are so sharp that they penetrate the bark, and their forcipate hands (Chapter 1), which allow the koala to grip the trunk. Smooth-barked trees that are often climbed can be indentified by numerous puncture marks on the trunk. Animals may pause during a climb, squatting on the hindlimbs and supporting the body with the forelimbs (Figs. 5.3 and 5.4). Koalas always walk down the trunk in an upright posture, extending the limbs in the reverse order to that adopted when walking up the trunk.

Before climbing, they usually pause to sniff the base of the tree. Adult males may mark the base with secretions from the sternal gland and animals of either sex may dribble urine against the trunk and on the ground close to the trunk (see below). Then they either walk slowly or bound up the tree. Koalas walking up a tree extend one forelimb, then the opposing hindlimb, the other forelimb and finally the remaining hindlimb. When bounding up the trunk, the animal leans back on the rump and hindlimbs and then, with a quick spring, thrusts the body and forelimbs upwards. The claws on the digits are dug into the trunk, the body is pulled upwards and the hindlimbs brought into a tucked position

Once near the canopy, koalas walk head first along the top of branches, passing the hands over one another and the feet over one another. The first two digits of each hand grasp one side of the branch and the remaining three digits the other. The foot is long and is oriented towards the side so that the clawless big toe points in the direction the animal is moving. The feet are cupped over the branch with the big toe grasping one side of the branch and the remaining toes the other.

**Figure 5.3**
A koala walking up a tree (*photograph by Peter Fell*).

# CHAPTER FIVE

**Figure 5.4**
Pausing during a climb up a tree. The body is held to the tree by the extended forelimbs and supported on the flexed hindlimbs. The white box on the throat houses a radio transmitter which is used to locate the animal (*photograph by David Curl*).

Koalas frequently stretch to cross narrow gaps between branches, and occasionally jump across gaps by releasing the forelimbs, extending the hindlimbs and jumping to grasp their target with the forepaws. Ronald Strahan (1978) has suggested that the stocky trunk and disproportionately long limbs of the koalas are adaptations to leaping in a more or less upright posture.

Feeding tends to occur towards the end of branches where the foliage of eucalypts is concentrated. Here koalas either sit or stand, clasping the branch with the feet and one forepaw while reaching to clasp stems with the other forepaw. In this way they draw individual stems towards the mouth. At the beginning of a bout of feeding they often sniff leaves at the tip of each stem and reject entire stems of leaves. Later, if most stems are accepted, they become more cavalier in their approach. Stems with leaves are manipulated with the incisors and forepaws, until the leaves are in a position where they can be bitten off (Chapter 3). Buds are removed from branches in the same manner as leaves, but bark is stripped from small stems with the incisors.

Small animals move to the smallest branches to feed, but large animals, especially large males, are very wary of moving to thin branches and often feed by pulling and breaking branches up to 2 centimetres in diameter. Occasionally branches do break under their weight and the animals fall, but injuries resulting from falls appear infrequent.

A surprising discovery of studies of the activity of koalas is how little time they spend grooming. Malcolm Smith (1979) watched his captive animals over two-hour periods, and found that on average, animals groomed only seven times during each period. Furthermore, most of these instances of grooming involved only a single scratch. Eighty-eight per cent of the grooming he observed involved the syndactyl second and third toes of the foot (the so-called toilet claws). These toes serve as a comb, and hair is sometimes found in the gap between them. Smith also observed koalas scratching with the claws of the forepaws and nibbling the fur with the incisors. Foot scratching was directed over most of the body, hand scratching was directed at the hindlimbs, feet, rump and lower flanks, and the animals nibbled the hindlimbs and feet. Although koalas appear to groom infrequently, it is apparently sufficient to keep the fur free from deadlocks. Old and debilitated koalas that have stopped grooming have deadlocks of thick and matted fur on the rump and flanks. It is a reflection on the knowledge of our politicians that a Minister for Tourism once decried the image of the koala as Australia's principal tourist attraction by describing koalas as 'flea ridden'. However examination of the pelt reveals that they are remarkably free of ectoparasites, and this may be the reason why they groom so little.

## 5.3 SOCIAL BEHAVIOUR

Gerald Durrell once described the koala as the most boring mammal he has laid eyes on. Most casual observers would probably agree and wonder whether the koala possesses any social behaviour. They are very solitary animals and, with the exception of mothers and their young, are usually found alone in a tree. Peter Mitchell (1988) found that slightly in excess of 93 per cent of the animals he sighted were either animals alone in a tree or mothers with young on their backs. With this low rate of interaction it is easy to see why the investigation of the social behaviour of koalas requires patient observation.

The earliest study of social behaviour in a natural population of koalas was by Ian Eberhard (1972, 1978). He concluded that adult koalas in the population on Kangaroo Island, South Australia, were sedentary, with each koala using about 15 trees in a home range of about 1 to 2.5 hectares. The ranges of individuals appeared to be largely separate, although he did observe some overlap of the ranges of males and females and sharing of some trees. This suggested that koalas might defend some type of territory against other koalas, but how this was achieved was not clear.

A more thorough investigation has been made by Peter Mitchell (1988) of a particularly dense population on French Island. Here the koalas were found in clusters scattered through a woodland, with each cluster associated with a patch of large trees. The clusters consisted of two to six females and a number of males. The females had home ranges of about 1 hectare, and these ranges showed some overlap (Fig.5.5). The males also had ranges that overlapped with other males and with females. Some, usually the older and larger males, had home ranges of about 1.5 hectares, but others had home ranges which were similar in size to the female ranges. Mitchell found no evidence of territoriality.

Male koalas, particularly the older and larger males, moved more often than females, and so encountered other animals more frequently. Encounters between males were

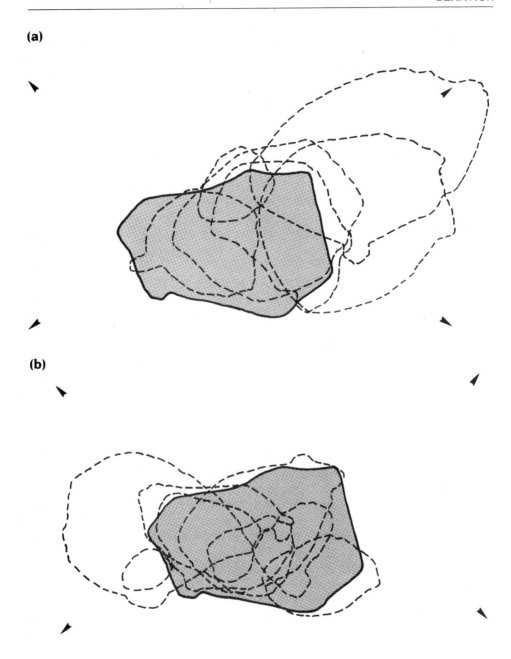

**Figure 5.5**
Home range boundaries of a cluster of koalas on French Island. (**a**) All males including the alpha male (*solid line*). (**b**) The alpha male (*solid line*) and all females. The home ranges of males are generally larger than those of females and more dispersed. The home range of the alpha male overlaps the range of all females. Arrow heads identify the location of grid reference points surrounding the cluster.

sometimes aggressive. In these encounters, one animal — usually the one entering the tree — would rush up to attack the other animal. The other animal either retreated to the end of the branch, or raced past the attacker and out of the tree. Sometimes animals at the end of a branch would try to leave the tree, but would be chased back by the attacker sitting on the same branch. If the attacker managed to reach the other animal, it would thrust one arm around its shoulders and grasp the elbow with its teeth, so holding the other animal, or even pulling it from the tree. Deep wounds inflicted in this way were seen on the elbows of several males. If the other animal left the tree, it was usually only chased a few metres before the attacker returned to the tree, where it often bellowed and marked the trunk with secretions of the sternal gland. During some encounters, the male resident in the tree would quietly retreat to the end of a branch as the intruder entered the tree, and was sometimes not noticed or was ignored by the intruder.

After observing many aggressive encounters, Mitchell was able to show that there was a dominance hierarchy among males. Males high up in the hierarchy had larger home ranges than those near the bottom. In the clusters of animals, one male, the alpha male, was clearly dominant over all other males he encountered. Occasionally, males made excursions into areas used by other clusters. Observations of these events showed one alpha male was dominant over others, even those outside his normal home range.

When adult males approached and entered trees occupied by females, the females would give low snarls or weak bellows and sometimes retreated. As the male came closer, the calls became louder and longer. If the male attacked, as often happened with adult males, he would attempt to hold the female between himself and the branch, sometimes upside down, and grasp the fur of the female's head or neck with his teeth. If the female was in the right position for mating, pelvic thrusting would commence. The usual response of the female was to struggle away from the male, striking or biting his head and squawking and screaming loudly. After such an attack, the male usually backed down the tree, bellowing as he went. Most males stayed in the tree and tried to mate a second or even third time. Others immediately left the tree, and occasionally returned to try again.

If one of the alpha males heard the sound of an attempt at mating nearby, he immediately left his tree and approached the sound, sometimes at a run. If the other male was still in

**Figure 5.6**
A male koala fending off an approaching male. Notice the ears of the defending male pressed against the head (*photograph by Peter Fell*).

the tree with the female, he was immediately attacked and evicted from the tree. The alpha male then remained in the tree occupied by the female and sometimes stayed nearby her for several hours.

Complete matings were seen on only a few occasions and always involved alpha males. They lasted for only one to two minutes. The male mounted the female, pulled her head back and gained intromission from behind with a series of pelvic thrusts. After a short bout of thrusting, both the male and female showed conspicuous contractions of the abdomen. The female then forced herself free of the male, giving low squawks, while the male moved down the tree, bellowing as he went.

These observations suggest that male aggression is associated with competition for mating partners, with the alpha males having greater access than other males. We are as yet uncertain when males first attempt to assert dominance, but it probably occurs at about five years of age when they are close to their maximum size. Males may not live much beyond 10 years in natural populations so that their reign, if they achieve alpha status, may last no more than five years.

Whether females select between males for mates is also uncertain. The screams and wails associated with attempted matings increase the chance that an alpha male will interrupt other males and try themselves to mate the female. This does not explain why males, particularly males low in the dominance hierarchy, try to mate with females that are not in oestrus, especially when they are at great risk of either falling from the tree or being attacked by the alpha male. This may be a form of courtship, but there is no evidence that females can distinguish between individuals, or that the courtship affects the response of females or the success of males. There is still much to be learnt about the mating system of the koala.

## 5.4 INFANT–PARENT BEHAVIOUR

The only social interaction between koalas that can be readily observed is the long and close association between mother and young. But even this interaction involves little interplay of behaviour (Fig. 5.7).

Mothers show no obvious behaviour towards young while they are in the pouch, and make no effort to clean the pouch. We have even found females with mummified pouch

BEHAVIOR

**Figure 5.7**
A mother with a seven month old back young. Other close associations between koalas are very infrequent (*photograph by Peter Fell*).

young which are covered with a heavy, brown, tar-like secretion.

Only a slight increase in behavioural interplay follows the departure of the young from the pouch. There is no obvious behavioural faciliation by the mother of the young's attempts to feed on pap, but she may spread her legs and so stretch the entrance of the pouch when small cubs are entering the pouch. The position cubs assume on the mother's back or belly appears to be entirely of the cub's choice, and the cub may even have to force its way between the forelimbs of the mother to adopt a sleeping posture on the belly. Mothers occasionally lick cubs when they are climbing over them, but this does not appear to be a consistent element in their behaviour.

Cubs removed from their mother emit a repeated, high-pitched squeak which elicits a searching response from the mother. Squeaking ceases as soon as the cubs climbs on to the mother's back. Occasionally cubs climb on to the backs of other females and even males, and are carried and tolerated by these animals. This occurs commonly in zoos where mothers may inadvertently swap young and seemingly do not recognise the change. Occasionally this happens in the wild and there it can lead to mortality of young cubs if they choose a male, or a female which is not lactating.

Clearly the bond between mother and infant in the koala is simple, and the mother's role in maintaining this bond is largely passive. Malcolm Smith (1979) suggested that the simplicity of the bond may be due to their solitary nature, which in itself greatly reduces the dangers of misidentification. So why does the association between mother and young last for so long? Part of the answer may lie in the threat of predation. Cubs, by virtue of their size, are prone to capture by predators such as the powerful owl. By clinging to the mother their small size may be disguised. But Smith also noted that the development of the manipulative skills which are required to eat leaves is slow, and that it is only towards the end of the period of dependence that these skills are fully developed. Perhaps it is the slow development of these skills, and the slow development of the digestive tract, that prolongs the cub's dependence on lactation as a supplementary source of nourishment, and so its dependence on the mother.

## 5.5 COMMUNICATION

Koalas have two modes of communication, scent marking and vocalising, which do not involve direct confrontation. They employ two modes of scent marking. One of these involves use of the sternal gland of males. This gland is located in the centre of the chest and is marked by a streak of naked skin, surrounded by hair stained with a pungent, oily, orange-coloured secretion (Frontispiece). Males mark the trees they enter by rubbing the sternal gland against the base of the tree and against the trunk and knobs as they climb (Fig.5.8). To mark the base of the tree, they flatten the

**Figure 5.8**
A male koala scent marking a branch while climbing (*photograph by* Peter Fell).

chest against the trunk and rub it up and down about six times. The second mode of scent marking involves the use of urine. Both sexes occasionally dribble urine on the trunk and sometimes on the ground close to the tree. The odoriferous component of the urine is probably derived from the paracloacal glands, two small glands which open into the urogenital sinus.

The precise function of scent marking is not clear. Animals entering trees sniff the base and other marked sites, and occasionally move to another tree. Perhaps it serves to warn off other individuals from a tree occupied by the marker. It may also serve to warn subordinate males of the presence of a dominant male, or indicate the presence and perhaps reproductive status of females.

The vocalisations of koalas are more diverse than scent marks in both type and message they convey. Malcolm Smith (1980) identified eight different vocalisations, but acknowledged that there were no distinct boundaries between some of these.

The most extraordinary of the vocalisations is the bellow. It consists of a continuous series of calls, each call made up of two phases. The first is a tremulous inhalation, which Smith likened to a 'snore', and is followed immediately by the second phase, a shorter, louder, tremulous, exhaled 'belch'. Within each bellow the calls rise to a crescendo and then fade. The bellow usually lasts about 10 to 30 seconds when fully expressed, and sometimes several bellows are emitted in succession. Most bellows appear to be emitted spontaneously but some are given in reply to the bellows of other males. Males also bellow after aggressive encounters in which they have been the victor, and after attempts at mating. The frequency of bellowing increases during the mating season and the bellows may increase in loudness. There is no overt response to another individual's bellows other than sometimes bellowing in reply. Nevertheless bellows probably communicate the presence of an individual, and perhaps its status. Females occasionally bellow, and this usually occurs when another individual is about to enter the tree they are occupying. The bellow is low pitched and is therefore suited for communication over distance. On still nights it is possible to hear a male calling some 800 metres away. There is some circumstantial evidence that the bellows of males may also serve to attract females in sparse populations. Several years ago we released a number of koalas and radio-tracked them as they dispersed from the release site. One female settled in an isolated patch of manna gum where she remained until the beginning of the breeding season, 10 months later. She was then found 2 kilometres to the north, in the vicinity of two males. Six weeks later she was back in the manna gum with a newly attached pouch young. This was conceived at the time the female was found with the males. The most likely means of locating these males would have been to trace their calls.

Other vocalisations of the koala clearly have different functions. We have already seen how the repeated squeak of cubs serves to attract adults to their presence. The wails, snarls and screams of females probably serve as a defensive threat.

The overall impression gained from their range of social

signals is that most serve to preserve the social distance of an animal with a simple solitary social system. The limited behavioural repertoire required to maintain this social system and to acquire food is consistent with what we learnt earlier of their brain — small with relatively poor development of areas controlling movement.

# CHAPTER 6

# PREHISTORY AND HISTORY

### 6.1 ABORIGINES AND THE KOALA

To the Aborigines of eastern Australia, the koala was a source of food, a subject of mythology and an object of respect. An appreciation of these values can be gained from early historical records and from accounts of Aboriginal myths.

A substantial proportion of the Aboriginal population of eastern Australia lived close to the coast and along inland rivers, areas which are favoured by the koala today. For these people the koala was a substantial and easily captured food item. Most were caught by climbing trees and chasing the koala to the very extremity of a branch where it was either caught alive or killed with a stone axe. A more ingenious mode of capture was described by William Govatt in an early account of the natural history of the koala (see Iredale and Whitley 1934). This involved the use of a loop of ropy stringybark tied to the end of a long pole. The loop was juggled over the head of the koala and the pole was immediately twisted to tighten the noose. The koala was then pulled down the tree. The subsequent preparation of the koala depended largely upon myths and rituals.

Aborigines recognise special bonds with particular species of animals, and the koala is among these. In some tribes this was expressed as a totemism involving a bond between a clan and a species. Individuals of a clan believed their lives to be intimately linked with the life of the totem animal. Membership of a totem was inherited. Some of the tribes of south-eastern Australia had a different concept in which one or other sex of a tribe was linked to a species. As with totemism, the lives of individuals of that sex were bound to the lives of individuals of the species. The 'sex animals' could not be killed, for to do so would be killing one of the tribe, and potentially the life of the animal to which your own life was bound. Possibly because of its value as a source of food, the koala was not a sex animal, but tribes that adopted the 'sex animal' concept also held other animals in special esteem, and the koala was among these. The Kurnai tribes of Gippsland had upwards of 20 animals accorded special status and referred to these as Mukkurnai. Muk-kurnai were believed to be their progenitors. It was not taboo to eat Muk-kurnai, and their flesh, known as Muk-jak, was also held in special esteem.

The Kulin tribes of southern central Victoria did not grant the koala esteemed status, but nevertheless had certain rituals when roasting koalas which were intended to convey their respect. These rituals were considered by Aldo Massola (1968) to have origins similar to the rituals surrounding the Kurnai's esteemed animals. One of these rituals was the habit of roasting koalas with their skin on. Massola describes two myths which maintained this practice. The first is the myth of Kurburu, the koala. It tells of the Kulin preparing the feast on the gum of acacias, which they were mixing with water in wooden bowls called tarnuks, when Kurburu wandered by. Kurburu asked them for some of the gum but they suggested he gather some for himself. This annoyed Kurburu and while the Kulin were gathering more gum, he placed their tarnuks full of water at the top of a young tree. Kurburu then climbed the tree and perched in a fork near the top. He urged the tree to grow and soon it was taller than any other tree.

On their return, the Kulin searched for their tarnuks, eventually spotting the tarnuks and Kurburu at the top of the tree. They shouted to Kurburu to return the tarnuks, but he chose to ignore them. One of the men began to climb the tree intent on killing Kurburu, but as he approached, Kurburu picked up a tarnuk full of water and hurled it at the man, who lost his footing and fell to his death. Others climbed the tree but suffered a similar fate.

The smell of the dead attracted Bunjil, the Father of the Kulin, and on reaching the tree, he called to Kurburu to give him a drink from the tarnuks. But Kurburu told Bunjil to climb the tree to fetch his own drink. This angered Bunjil

# CHAPTER SIX

who expected everyone to obey him. He instructed two of his men in the art of avoiding the tarnuks as they climbed the tree. He then sent them up the tree, one either side. When Kurburu heard them coming he hurled a tarnuk in their direction. One of the young men leaned to one side and the tarnuk passed under his arm. Other tarnuks were avoided in the same way, until Kurburu, frightened by their approach, abandoned the tarnuks and climbed higher. After satisfying their thirst the young men killed Kurburu. Bunjil then ordered his men to come down and bring the remaining tarnuks. These he returned to the Kulin, instructing them always to cook koalas with their skin on, and break their legs so that they would be unable to steal their tarnuks.

The second myth describes how the Kulin once skinned koalas before cooking them, a practice to which the koalas objected. One day when the people were away from their camp, the koalas gathered all of the tarnuks and hid them. They also drained all of the water holes. The people returned to find that they had no water and cried in their thirst.

Their cries attracted Karakarook, the woman, from the sky. She listened to the plight of the Kulin and then to the resentment of the koalas. To settle the dispute, Karakarook permitted the Kulin to kill and eat koalas, provided they showed koalas respect by leaving the skin on. The koalas were never to steal the tarnuks and were to offer the Kulin sound advice.

The Kulins' insistence in leaving the skin on the koala is surprising as it denied them pelts whose insulative properties are superior to those of any other marsupial (Chapter 5). Nevertheless skinning a koala under any circumstances was clearly taboo. Massola relates an account of an Assistant Protector of Aborigines for Victoria in which he described the difficulties he faced when persuading a man to skin a koala so that he could use the pelt for a hat. After skinning the koala the Aborigine became very excited, claiming that the tribe would lose all of its supplies of water. The 'old doctors' joined the fracas, and on their insistence the skin was returned and buried with the half baked carcass in a manner befitting a black man. Only in this way did they believe that the koalas would be appeased and their supplies of water secured.

The fate of the carcasses of captured koalas was also guided by ritual. A.W. Howitt (1904) described how the Kurnai either cooked koalas where they were caught, or carried them back to the camp raw, according to the distance. If only one koala was killed it was given to the wife's parents, sometimes after the hunter removed and kept the liver for himself and wife. If two were caught, the second was given to the parents of the hunter. If three were caught, two went to the wives' parents and one to the hunters' parents. If the wife had no food, her parents would offer her some of the carcass, but on the morning after the koalas were captured.

Even the division of the carcass was ritualised. This depended upon the family structure and who among the relatives was present in the camp. Howitt gives as an example the division of the carcass when the young man's parents were present: the father received the right hindlimb, the mother the left hindlimb, the eldest brother the right forelimb, the younger brother the left forelimb; the elder sister received the backbone, the younger sister the liver. The right ribs were given to a paternal uncle, a piece of the flank to a maternal uncle, and the head went to the young men's camp. The hunter received the left ribs.

Other myths and legends had no direct bearing on how Aborigines handled koalas, but show the Aborigines' awareness of some of the unusual features of the koala, such as the absence of a tail. For example, a Kurnai myth, set at a time of a severe drought, tells of how the only person who did not suffer was Gula, the koala, who in those days possessed a tail.

Gula was suspected of having a secret supply of water but no one could find it. Eventually the tribe enlisted the help of Buln-buln, the lyre-bird. He followed Gula and watched him climb a tree and hang from a branch by his tail to drink from water in a tree hollow. Buln-buln set fire to the tree, which burst in the heat, showering water in all directions for the people to drink. Gula jumped to escape the flames, but in doing so left his tail coiled around the branch. That is why the koala has no tail.

Perhaps the most extraordinary illustration of the relationship between Aborigines and the koala is the account of William Thomas, an Assistant Protector of Aborigines, who was tracking five men with the assistance of a Kulin tracker. After losing the trail, they were attracted by the noise of a koala in a tree they were passing. A parley followed between the tracker and the koala, at the conclusion of which the tracker admitted his foolishness, set off in a new direction and, within one and a half miles, came once more upon the tracks of the missing men (Massola 1968).

## 6.2 DISCOVERY AND EXPLOITATION BY EUROPEANS

One of the most curious aspects of the history of the koala is that it escaped discovery by Captain Cook and his crew, and by the first settlers. In fact the first mention in historical records appears in the journal of John Price, a servant of Governor Hunter, in an account of an excursion south-west of Sydney in January 1798. Price mentions an animal, 'which the natives called a cullawine, which resembles the sloths in America', (Iredale and Whitley 1934, p. 62). The first specimen was not collected until four years later when Ensign F. Barrallier sent the feet of a dismembered koala to Governor King. Barrallier referred to the collection of this specimen in his journal: 'Gory told me that they had brought portions of a monkey (in the native language Colo), but they had cut it into pieces, and the head, which I should have liked to secure, had disappeared. I could only get two feet through an exchange which Gory made for two spears and one tomahawk. I sent these two feet to the Governor in a bottle of brandy' (Iredale and Whitley 1934, p.62). The first live animal was obtained by Barrallier the following year.

The belated discovery of the koala does not indicate reticence on the part of the early settlers to collect the native fauna. By 1803 Europeans were already familiar with the eastern grey kangaroo, quokka, tammar, common ringtail possum, eastern native cat, thylacine and banded hare wallaby. Rather it suggests that the density of koalas was low at the time of European settlement. This view is supported by the records of settlers and explorers in other parts of the koala's range.

Bob Warneke (1978) attempted to assess the distribution and status of the koala south of the Murray River at the time of European settlement by searching a number of historical accounts for mention of native fauna. He found very few references to koalas, and all sightings were from the dense forests of eastern Victoria. Warneke attached particular significance to this, noting that koalas are more readily seen in the open forests that are their favoured habitat today than in dense forest. It suggests that the koala was very rare or absent from open forests. Harry Parris (1948) found no mention of koalas in historical records of the lower Goulburn district of Victoria when it was first settled, but some were seen 15 years later. John Gould (1863) made substantial collections in eastern New South Wales and south-western Queensland between 1838 and 1840, with the assistance of Aborigines, but was only able to obtain koalas after a diligent search. He commented on the very patchy distribution of the koala, and expressed concern for its future. As Warneke points out, Gould's comments cannot be lightly dismissed, as he was a skilled and patient collector and searched well beyond the limits of settlement.

Harry Parris was probably the first to suggest that the rarity of koalas at the time of European settlement was due to predation by Aborigines, noting the increase in abundance of koalas as the Aboriginal population declined. This explanation would also account for the early sightings of koalas in dense forest, since Aborigines spent little time hunting there. The influence of Aboriginal hunting was probably compounded by predation by dingoes and by mortality resulting from fires.

The impact of this predation can be gauged by the changes in the abundance of koalas which followed the decline of the Aborigines. George Robinson, the Chief Protector of Aborigines in Victoria, travelled through south Gippsland in 1844, and commented (Mackaness 1978, p. 8): 'The forest animals have vastly increased in abundance since the destruction of the local inhabitants (aborigines) ... the Yowenjee, a powerful section of the Boonwerong nation, who have (with the exception of two individuals) been exterminated by the Gipps Land aborigines. The Phascolarctos Fuscus, Carbora of the natives, Bear or Monkey of the settlers ... were in places abundant'. Parris observed that there were 'thousands' of koalas in the red gum along the Goulburn River by 1870 where none were seen 30 years earlier. And some of the first settlers in south Gippsland, arriving some 40 years after the demise of local tribes, wrote that koalas were so abundant that their grunting kept them awake at night.

The rapid growth of populations seems to have occurred throughout the range and, if judged by the number of skins sold to the fur trade, koalas reached extraordinary abundance. But by the turn of the century the slaughter of koalas for their pelts had had a profound effect on their abundance and range.

This slaughter was horrendous judging from the few records that are available to us. The most reliable information comes from records for the last two open seasons in Queensland where one million 'bear skins' were marketed as a result of a six-month season involving ten thousand licensed trappers in 1919, and 584 738 skins from a one-month season in 1927. The first of these figures is even more extraordinary when it is realised that the 1919 open season was preceded by another open season only 18 months previously. Further south the skin trade had already dramatically reduced the numbers of koalas, yet in 1908, 57 933 skins were traded through the Sydney market.

Public concern for the future of the koala ultimately led to legislation protecting it. In Victoria, this was achieved under a proclamation in 1898 with the confusing title 'Native bears to be deemed native game and protected'. New South Wales followed with legislation providing for closed seasons on the hunting of koalas in 1903, and absolute protection in 1909. The *Native Animals Protection Act* of 1906 in Queensland provided for a closed season from 1 November to 30 April, but with the opportunity for hunting during open seasons. Six open seasons were ultimately declared but this practice was abandoned as a result of public outcry after the 1927 season. However, even with the imposition of legislation, large numbers of skins were still traded and sold as wombat.

Banning of hunting did not lead to a spectacular rise in the numbers of koalas, and indeed flimsy evidence suggests that numbers continued to decline until about 1920 in Victoria and the early 1930s in Queensland. It was estimated that there were 500 to 1000 koalas in Victoria in the early 1920s, although this was a rough estimate rather than a figure based upon surveys. Indeed there has been no systematic survey of koala numbers to the present.

Several causes can be found for their purported decline. Clearing of woodland and open forest has substantially reduced habitat, and the catastrophic fires which accompanied European presence had a profound influence on their abundance. Koalas were very common in south Gippsland in 1898, but they were not seen in the area for several decades after a severe fire in 1900. Although clearing and fires are acknowledged to have had their effect, most observers have implicated disease as the cause of the slow recovery.

The last 30 to 40 years have seen some recovery to the point where, in southern Australia, it has been necessary to remove substantial numbers of animals from some sites where they were overbrowsing and endangering their food trees. Nevertheless there remains considerable public concern for their future, especially in northern New South Wales and Queensland.

# CHAPTER 7

# CONSERVATION AND MANAGEMENT

## 7.1 DISEASE

The notion that disease may influence the numbers of koalas was first publicised by Ambrose Pratt (1937), when he referred to the opinion of Noel Burnett that the condition of cystic reproductive organs would inevitably wipe out the koala, leading to its extinction within a few years. Ellis Troughton (1941) was presumably referring to the reproductive tract when he suggested that koalas were subject to a variety of diseases due to their 'peculiar anatomy' (see Chapter 3), although he only identified ophthalmic disease, a disease of the eyes. This he suggested was introduced by Europeans and had killed large numbers of koalas during epidemics in 1887-89 and 1900-03. Recently Steven Brown and Frank Carrick (1985) aroused public concern when they claimed that a disease epidemic, which caused blindness, infertility and death, was sweeping through koala populations, and warned that extinction was a potential outcome.

The prevalance of diseases of the reproductive tract has impressed most biologists who have worked with koalas. Probably the first to draw attention to this was the English anatomist J.P. Hill whose observations were published by C.H. O'Donoghue (1916). Hill noted that among the hundreds of reproductive tracts of marsupials he had examined, 'cystic ovaries' were only encountered in the koala, and that this condition was by no means uncommon. He also observed that the cystic condition was not confined to the ovaries but also occurred in the fallopian tubes and uterine horns. Hill's reference to 'cystic ovaries' was unfortunate, for he was followed in this terminology, until recently, by almost every biologist concerned with the health of koalas. Cystic ovary disease in humans is characterised by large cysts in the ovaries and is thought to be caused by hormonal imbalance. This is not the disease found in koalas.

Correct identification of the condition in koalas was established by David Obendorf (1981) following a rigorous examination of the pathology of the reproductive tracts of diseased koalas. Obendorf concluded that the ovaries in these animals were normal in appearance and showed signs of normal function. This has been confirmed recently by Kath Handasyde (1986), who found that the profiles of the ovarian hormones, oestradiol and progesterone, were identical in normal and diseased koalas. David Obendorf observed that the cysts that others had detected were the result of fluid accumulating in the ovarian bursae, fallopian tubes and uterine horns, leading to swelling of those parts of the reproductive tract. This accumulation was caused by infection obliterating the pore which communicates between the cavity of the ovarian bursae and the body cavity (Fig. 3.9). E.S. Finckh and A. Bolliger (1963) had earlier suggested that the small size of this pore predisposed koalas to the cystic condition. Obendorf found that the disease was characterised by infection at various levels in the vaginae, uterine horns, fallopian tubes and ovarian bursae, and concluded that it was the result of an infection progressing up the tract from the urogenital sinus. Identification of the organism responsible for the infection came in 1984 when teams led by Steven Brown and Ken McColl independently implicated the bacterium *Chlamydia psittaci*.

## CHLAMYDIA AND CHLAMYDIOSIS

The name *Chlamydia* comes from the Greek *chlamys*, a cloak, and this probably speaks for the frustration engendered in microbiologists when they first attempted to isolate and culture the micro-organism. *Chlamydia* grows and multiplies within the cells of its host and in this respect resembles a virus.

*Chlamydia* first attracted attention as a human pathogen which caused psittacosis or 'parrot fever'. As the name suggests, this disease was contracted from parrots, and its most severe expression was fatal pneumonia among the elderly. Scientific interest in *Chlamydia* was stimulated by fears of an epidemic of psittacosis in the United States and Europe in the 1930s. Concurrent studies on the natural hosts, parrots, in the United States and Australia showed that infection was widespread in both aviary and wild birds. It tended to be latent in mature birds, and only caused significant mortality in fledglings, and in adults exposed to the stress of captivity. Interest in the organism, by then identified as *Chlamydia psittaci*, began to wane when it became clear that it posed no great threat to mankind. Subsequently it has been recognised as one of the most ubiquitous parasites within the animal kingdom and its natural hosts include most birds and many mammals.

*Chlamydia* again attracted attention in the late 1970s when another species *C. trachomatis*, was perceived as a threat to human health. As there are striking similarities in the diseases caused by *C. trachomatis* in humans and *C. psittaci* in koalas, it is worthwhile investigating *C. trachomatis* further. As the name implies, *C. trachomatis* is responsible for the debilitating eye disease, trachoma, which has an equivalent in koalas referred to as 'pink eye' or keratoconjunctivitis. Trachoma is widespread in Third World communities living under conditions of poor hygiene. Being an adaptable organism, another strain of *C. trachomatis* has found a niche in affluent societies where it is now responsible for one of the more prevalent sexually transmitted diseases. Among human males the infection is confined to the lower urinary tract and causes little more than a burning sensation. Human females are often without symptoms, but the consequences are potentially serious. In about 10 per cent of cases the infection ascends the reproductive tract and causes severe pathological changes to the uterus and fallopian tubes. This expression of the disease is called pelvic inflammatory disease and usually leaves the sufferer infertile.

While there are some differences in pathology, the cystic condition in koalas is remarkably similar to human pelvic inflammatory disease. In koalas it is caused by an ascending infection of *C. psittaci*, and is now generally referred to as reproductive tract disease. This disease, 'pink eye' and another condition know as 'dirty tail', are collectively referred to as chlamydiosis. Dirty tail or 'wet tail' as it is sometimes called, is a disease of the urinary tract which leads to frequent urination, and results in soiling of the hindquarters. Severe infections may cause damage to the kidneys.

Although the observations of J.P. Hill and others suggest

that chlamydiosis is prevalent in koala populations, the actual prevalence did not become apparent until a survey was undertaken by Steven Brown in 1978. In Queensland, Brown found that the majority of females in most of the populations he examined were suffering from chlamydiosis and were infertile. And on Phillip Island, where our attention had been drawn to low fertility among females, 27 out of 30 females examined showed evidence of disease of the reproductive tract. Subsequently we have found evidence of *Chlamydia* in four of the seven large populations of koalas in Victoria and South Australia. The only populations known to us which are *Chlamydia*-free occur on French Island and at Sandy Point, in Victoria, and on Kangaroo Island, South Australia. Does this high incidence of infection mean that the koala is threatened with extinction? This question can best be answered by examining the history of two populations.

In 1953, 32 koalas were released on Raymond Island in the Gippsland Lakes. These koalas were drawn from the Phillip Island population where low fertility at that time suggest that chlamydiosis was as prevalent as it is today. Despite the likelihood of a high level of infection among the founding group, the population on Raymond Island has flourished. Surveys in 1980 and 1985 found in excess of 170 koalas on both occasions, and although 86 per cent of the population was infected with *C. psittaci* in 1985, 40 per cent of females were accompanied by young.

The population in the Brisbane Ranges National Park was established with koalas translocated from Quail Island in 1944, Phillip Island in 1945, French Island in 1957 and Phillip Island again in 1977. The levels of fertility among the Phillip Island females in 1944 and 1977 suggests that these groups were infected with *Chlamydia* when translocated. This population has also thrived, despite the early introduction of the disease. Seventy-eight per cent of individuals in this population are infected, yet at the present time 40 per cent of the females breed each year and koalas are common throughout the 7000 hectare park. As yet we do not have an estimate of the death rate in this population, but field observations suggest that it is low. There may be mortality among cubs, but predation is insignificant and few animals die as a result of chlamydial infections. If we assume an annual mortality of 10 per cent, then the annual growth rate of the population will be in the order of 10 percent and the population will double in size every 6.9 years!

From these and other observations we have concluded that chlamydiosis may not pose a threat to the koala. We

## CONSERVATION AND MANAGEMENT

**Figure 7.1**
Catching a koala. A locking noose is placed over the head and around the neck with the aid of the extendable aluminium pole. The pole is then withdrawn and the rope leading from the noose is held under tension. The cloth flag, attached to the tip of the pole, is waved above the head to induce the koala to climb down the trunk (*photograph by Michael Coyne*).

have evidence that a proportion of infected animals continue to breed, and that, at least as far as Victorian populations are concerned, the prevalence of the more debilitating diseases of pink eye and dirty tail is extremely low.

## 7.2 ALIENATION OF HABITAT

It has been estimated that the proportion of the Australian land surface covered with native forests has shrunk from 15 to 5.5 per cent since 1800. Furthermore, clearing of forest has resulted in fragmentation and consequently, isolation of much of the surviving forest and its inhabitants. Today isolation and fragmentation may pose a greater threat to the survival of koalas than chlamydiosis. To comprehend this threat it is instructive to understand the dynamics of koala populations.

Peter Mitchell (1988) has recently drawn together the information we have on koala populations, and the following model is based upon his synthesis. Patches of favoured habitat, which are unoccupied or support a low density of koalas, gain young animals through immigration from nearby populations. Animals, particularly females, which are born in a patch tend to remain there while the density is low, but emigrate as the density increases. This emigration is usually restricted to animals between two and four years of age, and occurs among males more than among females. Animals older than four years which are resident in the patch tend to remain there, unless their food trees become severely defoliated. Once the population attains a certain, relatively high density, all young born in the patch emigrate, and any animals immigrating into the patch depart after only a short stay. At this density the trees show evidence of browsing but do not appear threatened with defoliation. Populations may remain at these high densities for a number of years, and perhaps indefinitely if there is no disturbance. As the residents die, their place is taken by the young of residents or by immigrants.

It was previously thought that there was a high mortality among emigrating young, but Mitchell was able to radio-track a number of these and found that their chance of surviving was high. Most spent several years wandering and were often found in tree species rarely used by adults. Ultimately they settled in patches of favoured habitat where the residents tolerated their presence.

Clearing of forest poses the greatest threat where it eliminates sites where dispersing animals can settle. Under these circumstances the animals are forced to live at exceptionally high densities in favoured patches, or reside where it is less favourable, to the possible detriment of their well-being. It is the exceptionally high densities of koalas in favoured patches which leads to the defoliation of the trees. This can be illustrated by reviewing the history of several populations.

Some years ago our attention was drawn to a population of koalas in a reserve at Walkerville, Victoria, where there was evidence that koalas were defoliating their food trees. This reserve of 2000 hectares was bounded on one side by the sea and on the other sides by land cleared for pasture. Within the reserve the koalas were concentrated in small patches of swamp gum within a forest of narrow-leafed peppermint and messmate. Over a three-year period, one of us (R.M.) observed most of the food trees (swamp gum) stripped of their foliage and killed. Over the same period the nearby narrow-leafed peppermint and messmate showed no evidence of browsing. As more and more swamp gum died, the density of koalas fell from 3 animals per hectare to 0.7 animals per hectare. Whereas 36 per cent of females bred in the first year of the study, only 11 per cent bred in the final year. Mortality increased, especially among old females, but most of the change in density resulted from dispersal of animals, particularly young animals, into the surrounding forest. Often animals disappeared for a time and then returned to feed on the few epicormic leaves remaining on the swamp gum. Presumably these koalas dispersed in search of other patches of swamp gum, but failed to find any. Clearing of the scrub surrounding the reserve had eliminated these patches. The fact that the collapse of this population occurred in the presence of two other species of *Eucalyptus* which showed no detectable damage to their foliage, once again illustrates the fastidiousness of koalas.

The consequences of restricting the number of sites where dispersing animals may settle is spectacularly demonstrated by the history of two populations on islands. A small number of koalas were introduced to French Island between 1870 and 1890 and by 1920 there were clear signs that they were stripping many of the manna gums of foliage. The abundance of koalas at that time was extraordinary. John McNally (1957) recorded that one settler claimed to have counted 2300 animals on an 8 kilometre stretch of the island's west coast. Ultimately, in the 1930s, defoliation of trees had reached such proportions that substantial num-

CHAPTER SEVEN

**Figure 7.2**
A koala sitting among severely defoliated trees at Sandy Point, Victoria.

bers of koalas died from starvation. The extent of the population crash was such that 25 years elapsed before it was necessary to remove animals to prevent defoliation.

Nearby Quail Island was stocked with 165 koalas from French Island between 1930 and 1933. Ten years later the numbers of koalas had grown to the point where they had killed most of their food trees, destroyed much of the remaining vegetation, and many of the koalas had died from starvation. Even so, 1349 survivors were removed to other sites.

These examples should serve to remind those wishing to conserve populations of koalas of the need to maintain avenues of dispersal or be prepared to manage the population below levels at which serious defoliation of the koala's food trees occurs.

## 7.3 KOALAS AND URBANISATION

Koalas appear to do poorly in urbanised areas. Urbanisation at Phillip Island in Victoria and at Avalon and Port Macquarie in New South Wales has been associated with declines in populations of koalas. A number of factors appear to have contributed. Clearing of forest has resulted in fragmentation and reduction of habitat. *Chlamydia* is prevalent in the populations on Phillip Island and at Port Macquarie and this has led to low fertility and some mortality resulting from dirty tail and pink eye. However the factor which appears to be tipping the scales at these sites is mortality resulting from injuries inflicted by cars and domestic dogs. The koala population on Phillip Island is currently estimated to contain about 300 individuals. Records kept by wildlife ranges on the island show that roughly 50 koalas have been found dead in each of the past three years. Of these deaths, approximately 60 per cent can be attributed to injuries inflicted by cars, and 6 per cent to injuries inflicted by domestic dogs. It is not surprising that this population, with a fertility of about 17 per cent, is in sharp decline.

## 7.4 MANAGEMENT

The capacity of koalas to destroy their food resources, and their vulnerability in the face of urbanisation, point to the need for management of koala populations. Since the 1920s, it has been the practice in Victoria to remove koalas from sites where their food trees show signs of defoliation, and release them in suitable habitat within the former range of the species. Care is taken to ensure that the release sites contain more than 4000 hectares of suitable habitat and have security of tenure and management. The number of koalas translocated in this way has been substantial with the result that the species is now re-established over much of its former range in Victoria where forest and woodland remain.

Although this practice appears a satisfactory means of

containing the growth of koala populations, the release of koalas into new habitats is not without problems. Some translocations appear to have failed: the animals disappeared following release and were never seen again. Another concern is that most of the coastal manna gum woodland is now occupied, so that the suitability of other forest types for koalas must now be considered. Finally, the highly fecund populations, which pose the greatest threat to their resources and therefore require most intense management, are *Chlamydia*-free. Release of animals from these populations at sites already occupied by infected animals poses a threat to their survival.

To find solutions to some of these problems, we translocated four groups of koalas, using animals from the *Chlamydia*-free and intensively managed population on French Island. Each animal was fitted with a radio collar, allowing us to track and relocate them using a receiver fitted with a directional aerial (Fig. 7.3)

Almost all of the animals we translocated dispersed immediately after they were released and were widely spaced through the forest by the time they had settled two months later. Yet despite their spacing, all females which were capable of breeding, did so in the next two breeding seasons. We have already referred to the role the males' bellows may play in drawing together potential mates (Chapter 5). The strong tendency to disperse from the release site may explain why some translocations appear to have failed, but clearly this does not limit the potential of the released animals to establish a population.

Familiarity among the released koalas does not influence their tendency to disperse or settle together. Animals from the same cluster of trees on French Island showed no greater tendency to stay together than animals drawn from widely scattered points on the island. Release into an unfamiliar tree species may increase the tendency to disperse, even among females which are normally more sedentary than males. However, many of the animals settled in species totally unfamiliar to them. Our animals were drawn from manna gum and at one site were released into swamp gum, which also occurs on French Island and is a commonly used food tree in southern Victoria. Some settled in manna gum; some used manna gum and other eucalypts, such as long-leafed box, with which they were not familiar; and some settled where they only had access to unfamiliar species, However, in every instance they chose species which are known to be commonly used by koalas. There was no difference in the health or later reproduction of animals

**Figure 7.3**
Animals fitted with radio collars (see Fig. 5.4) are tracked using a portable receiver and a directional aerial (*photograph by Michael Coyne*).

settling in familiar or unfamiliar species, so it is clear that koalas can be successfully released into forests where the species of *Eucalyptus* are unfamiliar to the individual but are food species favoured by koalas. This observation allows us to confidently extend the range of forest types suitable for releases.

The most disturbing effects of these translocations resulted from an inadvertent release of animals into an area where, unknown to us, there was a sparse population of animals infected with *Chlamydia*. Those animals that contracted an infection failed to reproduce and some died. We suspect that animals from *Chlamydia*-free populations are more prone to reproductive failure and mortality following infection than animals drawn from infected populations. This poses a problem for which we have no immediate solution. The populations which require intense management today are mostly populations which have had no history of chlamydial infections. Many of the sites which are available for the release of animals have a low density of koalas which are infected with *Chlamydia*. Either we need to identify sites where there are no infected animals or find some means of protecting the released koalas.

## 7.5 THE FUTURE

Rather than concerning ourselves with the unlikely possibility that koalas may be declining to extinction, we see that the most pressing problem is to resolve what to do with koalas removed from overstocked areas. There is only a finite amount of habitat suitable for koalas and their numbers will eventually need to be regulated if this habitat is to persist. In this respect, the almost universal presence of *Chlamydia* among koalas may be a blessing. Although some effects of chlamydial infections (pink eye and dirty tail) are debilitating, the incidence of these does not appear high. At present, infertility as a result of reproductive tract disease, bush fires and drought are the only natural constraints on the growth of populations of koalas. These constraints are delaying us from asking a most difficult question: What do we do when we have too many koalas?

# BIBLIOGRAPHY

Archer M (1976) 'Phascolarctid origins and the potential of the selenodont molar in the evolution of diprotodont marsupials'. *Memoirs of the Queensland Museum* vol 17 pp 367-71

Archer M (1978) 'Koalas (*Phascolarctos*) and their significance in marsupial evolution' in T J Bergin (ed) *The Koala. Proceedings of the Taronga Symposium on Koala Biology, Management and Medicine* Zoological Parks Board of NSW Sydney pp 20-8

Archer M (1984) 'On the importance of being a koala' in M Archer & G Clayton (eds) *Vertebrate Zoogeography and Evolution in Australasia* Hesperian Press Carlisle, Western Australia pp 809-15

Bensley B A (1903) 'On the evolution of the Australian Marsupialia; with remarks on the relationships of the marsupials in general' *Transactions of the Linnean Society London (Zoology)* vol 9 pp 83-217

Brown A S, Carrick F N (1985) 'Koala disease breakthrough' *Australian Natural History* vol 21 pp 314-17

Brown A S, Carrick F N, Gordon G & Reynolds K (1984) 'Diagnosis and epidemiology of an infertility disease in the female koala' *Veterinary Radiology* vol 25 pp 242-8

Cork S J (1986) 'Foliage of *Eucalyptus punctata* and the maintenance nitrogen requirements of koalas, *Phascolarctos cinereus*' *Australian Journal of Zoology* vol 34 pp 17-23

Cork S J, Hume I D, & Dawson T J (1983) 'Digestion and metabolism of a mature foliar diet (*Eucalyptus punctata*) by an arboreal marsupial, the koala (*Phascolarctos cinereus*)' *Journal of Comparative Physiology* B vol 153 pp 181-90

Cork S J & Hume I D (1983) 'Microbial digestion in the koala (*Phascolarctos cinereus*, Marsupialia) an arboreal folivore' *Journal of Comparative Physiology* B vol 152 pp 131-35

Cork S J & Warner A C I 1983) 'The passage of digestion markers through the gut of a folivorous marsupial, the koala *Phascolarctos cinereus*' *Journal of Comparative Physiology* B vol 152 pp 43-51

Degabriele R & Dawson T J (1979) 'Metabolism and heat balance in an arboreal marsupial, the koala *Phascolarctos cinereus*' *Journal of Comparative Physiology* B vol 134 pp 293-301

Eberhard I H (1972) 'Ecology of the koala, *Phascolarctos cinereus* (Goldfuss), on Flinders Chase, Kangaroo Island' Ph.D. Thesis The University of Adelaide South Australia

Eberhard I H (1978) 'Ecology of the koala, *Phascolarctos cinereus* (Goldfuss) Marsupialia: Phascolarctidiae, in Australia' in G G Montgomery (ed) *The Ecology of Arboreal Folivores*. Smithsonian Institution Press Washington D C pp 315-28

Eberhard I H, McNamara J, Pierce R J & Southwell I A (1976) 'Inges-

# BIBLIOGRAPHY

tion and excretion of *Eucalyptus punctata* D.C. and its essential oil by the koala *Phascolarctos cinereus* (Goldfuss)' *Australian Journal of Zoology* vol 23 pp 169-78

Finckh E S & Bolliger A (1963) 'Serous cystadenomata of the ovary in the koala' *Journal of Pathology and Bacteriology* vol 85 pp 526-8

Fleay D (1937) 'Observations on the koala in captivity. Successful rearing in Melbourne Zoo' *Australian Zoologist* vol 9 pp 68-80

Gould J (1863) *The Mammals of Australia* Taylor & Francis London

Haight J R & Nelson J E (1987) 'A brain that doesn't fit its skull; a comparative study of the brain and endocranium of the koala, *Phascolarctos cinereus*' in M Archer (ed) *Possums and Opossums: Studies in Evolution* Surrey Beatty Sydney (in press)

Handasyde K A (1986) 'Factors affecting reproduction in the female koala (*Phascolarctos cinereus*)' PhD Thesis Monash University Clayton Victoria

Hindell M A, Handasyde K A & Lee A K (1985) 'Tree species selection by free-ranging koala populations in Victoria' *Australian Wildlife Research* vol 12 pp 137-44

Hindell M A (1984) 'The feeding ecology of the koala, *Phascolarctos cinereus*, in a mixed *Eucalyptus* forest' MSc Thesis Monash University Clayton Victoria

Home E (1808) 'An account of some peculiarities in the anatomical structure of the wombat' *Philosophical transactions of the Royal Society of London* pp 308-12

Howitt A W (1904) *The Native Tribes of South-east Australia* Macmillan New York

Hughes R L (1965) 'Comparative morphology of spermatozoa from five marsupial families' *Australian Journal of Zoology* vol 13 pp 553-43

Hughes R L (1974) 'Morphological studies on implantation in marsupials' *Journal of Reproduction and Fertility* vol 39 pp 173-86

Iredale T & Whitley G P (1934) 'The early history of the koala' *Victorian Naturalist* vol 51 pp 62-72

Kirsch J A W (1968) 'Prodromus of the comparative serology of marsupials' *Nature* vol 217 pp 418-20

Lanyon J M & Sanson G D (1986) 'Koala (*Phascolarctos cinereus*) dentition and nutrition. I. Morphology and occlusion of cheek teeth'. II. Implications of tooth wear in nutrition' *Journal of Zoology London* A vol 209 pp 155-68, 169-81

McColl K A, Martin R W, Gleeson L J, Handasyde K A & Lee A K (1984) '*Chlamydia* infection and infertility in the female koala (*Phascolarctos cinereus*)' *Veterinary Record* vol 115 p 655

Mackaness G (1978) 'George Augustus Robinson's journey into south-eastern Australia, 1844' *Australian Historical Monographs* 19 Review Publications Dubbo

MacKenzie W C (1918) *The Gastro-intestinal Tract in Monotremes and Marsupials* Critchley Parker Melbourne

MacKenzie W C (1919) *The Genito-urinary System in Monotremes and Marsupials* Jenkin Buxton Melbourne

McNally J (1957) 'A field survey of a koala population' *Proceedings of the Royal Zoological Society of New South Wales* 1957 pp 18-27

Martin R W (1981) 'Age-specific fertility in three populations of the koala, *Phascolarctos cinereus* (Goldfuss) in Victoria' *Australian Wildlife Research* vol 8 pp 275-83

Martin R W (1985) 'Overbrowsing, and in decline of a population of the koala, *Phascolarctos cinereus*, in Victoria. I. Food preference and food tree defoliation. II. Population condition. III. Population dynamics' *Australian Wildlife Research* vol 12 pp 355-65, 367-75, 377-85

Martin R W & Lee A K (1984) 'The koala, *Phascolarctos cinereus* the largest marsupial folivore' in A P Smithe & I D Hume (eds) *Possums and Gliders* Australian Mammal Society Sydney pp 463-7

Massola A (1968) '*Bunjil's cave: myths, legends and superstitions of the Aborigines of south-east Australia*' Lansdowne Press Melbourne

Minchin K (1937) 'Notes on the weaning of a young koala (*Phascolarctos cinereus*)' *Records of the South Australian Museum* vol 6 pp 1-3

Mitchell P J (1988) 'Social organization of the Koala, *Phascolarctos cinereus*' PhD Thesis Monash University Clayton Victoria

Nagy K A & Martin R W (1985) 'Field metabolic rate, water flux, food consumption and time budget of koalas, *Phascolarctos cinereus* (Marsupialia: Phascolarctidae) in Victoria' *Australian Journal of Zoology* vol 33 pp 655-65

Obendorf D (1981) 'Pathology of the female reproductive tract in the koala, *Phascolarctos cinereus* (Goldfuss), from Victoria, Australia' *Journal of Wildlife Diseases* vol 17 pp 587-92

O'Donoghue C H (1916) 'On the corpora lutea and interstitial tissue of the ovary in the marsupialia' *Quarterly Journal of Microscopical Science* vol 61 pp 433-73

Parris H S (1948) 'Koalas on the lower Goulburn' *Victorian Naturalist* vol 64 pp 192-3

Pratt A (1937) *The Call of the Koala* Robertson & Mullens Melbourne

Robbins M & Russell E (1978) 'Observations on movements and feeding activity of the koala in a semi-natural situation' in Bergin op cit pp 29-41

Sharpe L L (1980) 'Behaviour of the koala *Phascolarctos cinereus* (Goldfuss)' Honours Thesis Monash University Clayton Victoria

Smith M T A (1979) 'Notes on reproduction and growth in the koala, *Phascolarctos cinereus* (Goldfuss)' *Australian Wildlife Research* vol 6 pp 5-12

Smith M T A (1979, 1980) 'Behaviour of the koala, *Phascolarctos cinereus* (Goldfuss), in captivity. I. Non-social behaviour. II. Parental and infantile behaviour. III Vocalisations. IV. Scent-marking. V. Sexual behaviour' *Australian Wildlife Research* vol 6 pp 177-28, 129-40, vol 7 pp 13-34, 35-40, 41-52

Sonntag C F (1921) 'The comparative anatomy of the koala (*Phascolarctos cinereus*) and Vulpine Phalanger (*Trichosurus vulpecula*)' *Proceedings of the Zoological Society* London 1921 pp 547-77

Sonntag C F (1922) 'On the myology and classification of the wombat, koala and phalangers' *Proceedings of the Zoological*

Society London 1922 pp 863-96
Southwell I A (1978) 'Essential oil content of koala food' in Bergin op cit pp 62-74
Strahan R (1978) 'What is a Koala?' in Bergin op cit pp 3-19
Strahan R & Martin R W (1982) 'The koala: little fact, much emotion' in R H Grose & W D L Ride (eds) *Species at Risk: Research in Australia* Australian Academy of Science Canberra pp 147-55
Thomas O (1888) 'Catalogue of the Marsupialia and Monotremata in the collection of the British Museum (Natural History)' British Museum London
Thomas O (1923) 'On some Queensland phalangeridae' Annals and Magazine of Natural History vol 9 pp 246-50
Troughton E leG (1935) '*Phascolarctos cinereus* victor' *Australian Naturalist* vol 9 p 139
Troughton E leG (1941) *Furred Animals of Australia* Angus and Robertson Sydney
Ullrey D E, Robinson P T & Whetter P A (1981) 'Composition of preferred and rejected browse offered to captive koalas, *Phascolarctos cinereus* (Marsupialia)' *Australian Journal of Zoology* vol 29 pp 839-46
Warneke R M (1978) 'The status of the koala in Victoria' in Bergin op cit pp 109-14
Winge H (1893) 'Fossil and living marsupials (Marsupialia) from Lagoa Santa, Minas Gerais, Brazil, with a review of the interrelationships of the marsupials' E *Museo Lundii* vol 2 pp 1-132
Wood-Jones F (1924) '*The Mammals of South Australia. Part II. The Bandicoots and the Herbivorous Marsupials (The syndactylous Didelphia)*' Government Printer Adelaide pp 133-270

# INDEX

*Page numbers in italics refer to illustrations*

Aborigines
    hunting of koalas, 60, 78, 79, 80, 81, 83
    myths and legends, 78–81
activity cycle, 61, *62*
age
    determination of, 43
aggression, 70, *71*, 72, 76
ancestry, 14–21, *17*

behaviour
    aggressive, 70, *71*, 72, 76
    development of, 56, 57, 58, 74
    feeding, 34, 37, 54–56, 57, 61, *62*, 67
    grooming, 13, 61, 67, 68
    maternal, 58, 72–74, *73*, 76
    play, 57, 58
    reproductive, 70, 71, 72
    social, 61, 62, 68–77
bellowing, 51, 70, 72, 76, 94
birth, 51, 52, 53
bladder, *49*, 50
brain
    case, 46
    size, 46, *47*, 58, 77
    structure, *47*, 48
breeding season, 51, 52, 59
bushfires, effects on koalas, 25, 60, 83, 84

caecum, 30, 41–43 *42*, 45
cerebellum, *47*, 48
cerebral lobes, *47*, 48
cheek pouches, 20
chlamydiosis, 59, 86–88, 90, 93, 96
*Chlamydia psittaci*, 86, 87, 98
*Chlamydia trachomatis*, 87
cineole, 29, 30
claws, 12, *13*, 65, 67
cloaca, 50
coat
    colour, 12, 57
    insulation, 63, 80
colon, 41, *42*, 43
conservation, 84, 90–96
copulation, 72
cyanide poisoning, 29, 30, 31

defoliation, 90–93, *92*
development, 52–58, *53, 55*
diet, 23, 25–33
digestion, 34–43
digits, 12, *13*, 63, *64*, 65
dingo, 60, 83
discovery, 82

disease
    dirty tail, 87, 90, 93
    ophthalmic, 85, 87, 90
    pink eye, 87, 90, 93
    reproductive tract, 85, 86, 87, 88, 89, 90, 96
dispersal, 58, 90, 91, 94
distribution
    fossil, 23, *24*
    present, *24*, 25
    Queensland, *24*, 25
    South Australia, *24*, 25
    Victoria, *24*, 25
    Western Australia, *24*
drought, 60

ear, 12, 14, 54
essential oils, 29, 30, 31
*Eucalyptus* species, 25–28, 29, 30, 31, 32, 33, 91, 94, 95, 96
eye, 12, 54, 85

fallopian tube, 48, *49*, 50, 86
feeding
    behaviour, 31–36, 37, 54–56, 57, 67, 74
    daily cycle, 61, 62
feet, 12, *13*, 21, 63, *64*, 65, 67
female
    head shape, *14*
    pouch, 14, *15*, 53, 54
    reproductive system, 48, *49*, 50
fertility, 85, 86, 87, 88, 91, 96
fighting, 70, *71*, 72
food
    mastication of 35, 36, 37–40, *38, 40*, 43, 44
    preferences, 25–28, 32
    seasonal changes in 26, 27, 28
    trees, 25–28, 32, 94, 96
forest red gum, 25, 26, 32
fossil record, 16–18, *17*, 24
fur
    colour, 12, 57
    insulative value, 63, 80
    texture, 12, 67
    trade, 83, 84

gastric gland, 20, 21
gestation, 52
grey gum, 27, 30, 32
grooming, 13, 63, 67, 68
growth, 54–58, *56*
gut, 20, 21, 34, 35, 41, *42*, 43, 74

101

# INDEX

habitat, 23–25, 33, 82, 92, 93, 94, 95, 96, 97
hand, 12, *13*, 18, 20, 21, 37, 63, *64*, 65, 67
home range, 28, 58, 68, *69*

keratoconjunctivitis, 87
*Koobor jimbarrattii*, 16, *17*
*Koobor notabilis*, 16, *17*

lactation, 51, 55, 57, *59*
legal protection, 84
*Litokoala kutjamarpensis*, 16, *17*
locomotion, 57, 63–67, *64*, *65*
longevity, 60

male
    head shape, *14*
    sternal gland, *10*, 14, 64, 74, 75
management, 93–96
manna gum, 25, 26, 27, 28, 29, 31, 32, 33, 76, 91, 94
mastication, 35–37, *38*, 39, *40*, 42, 43, 44, 45
mating, 51, 72
metabolic rate, 45, 46
mortality, 60, 74, 88, 90, 91, 93

neonate, *53*, 54, *55*
nose pad, 12

oestrous cycle, 51, 52
ovarian
    bursa, 48, *49*, 50, 86
    cysts, 85, 86
    hormones, 86
ovary, 48, *49*

pap, 54–56
pap feeding, 54–56, *59*
paracloacal glands, 75
*Perikoala palankarinnica*, 16, *17*
*Phascolarctos*, 16, *17*, 21
*Phascolarctos cinereus*, *17*, 21, 22
*Phascolarctos stirtoni*, 16, *17*
physical characteristics, 1, 12, 13, 14
placenta, 53
population
    ecology, 88, 90, 91
    on islands, 91, 92, 93

pouch
    young, 52–57, 73
    structure, 14, *15*, 20
predators, 60, 74, 83, 88

radio-tracking, *66*, 76, 90, 94, *95*
relationships, with other marsupials, *17*, 18–21
reproduction, 48–50, *49*, 51, 52, 53, 58, *59*, 72
reproductive system, 20, 48–50, *49*, 86
rhinarium, 12
river red gum, 25, 26, 32, 33

scent marking, 64, 70, 74–76
sexual dimorphism, 11, 12
size, 11, 12
sleeping
    postures, 62
    daily cycle, 61, *63*
sperm, *19*
statues, 25, 85, 96
sternal gland, *10*, 14, 64, 70, 74–76
subspecies, 21, 22
swamp gum, 26, 27, 28, 32, 33, 91, 94

tail, 12, 20, 21
taxonomy, 21, 22
teeth
    canine, *36*, 37
    cusps, 18, 19, 37, 38, *39*, 43, *44*
    eruption, 56, 57
    fossil, 16, 17
    function, 36–40, *37*, *38*, *40*, 43–45, 56, 57, 60
    incisor, 18, *36*, 37, 56, 57, 67
    molar, 18, 19, *36*, 37, *38*, *39*, *40*, 43, *44*, 57
    premolar, *36*, 37
temperature regulation, 63
toes, *13*, 18, 20, 63, *64*, 65, 67
tooth wear, 43, 44, 45

uterus, 48, *49*, 50, 86

vaginae, *49*, 50, 86
vocalisations, 70, 72, 74, 76, 94

water requirements, 46
weight, 11, 12, 13
wombats, 18, 19, 20, 21, 84